Minecraft™ Mastery: Build Your Own Redstone Contraptions and Mods

Matthew Monk
Simon Monk

New York   Chicago   San Francisco
Athens   London   Madrid   Mexico City
Milan   New Delhi   Singapore   Sydney   Toronto

Cataloging-in-Publication Data is on file with the Library of Congress

Minecraft™ Mastery: Build Your Own Redstone Contraptions and Mods

234567890   DOC DOC   10987654

ISBN   978-0-07-183966-2
MHID   0-07-183966-6

Sponsoring Editor	Acquisitions Coordinator	Production Supervisor
Roger Stewart	Amy Stonebraker	Jean Bodeaux
Editorial Supervisor	**Copy Editor**	**Composition**
Patty Mon	Margaret Berson	Cenveo Publisher Services
Project Manager	**Proofreader**	**Illustration**
Vastavikta Sharma, Cenveo® Publisher Services	Claire Splan	Cenveo Publisher Services
	Indexer	**Art Director, Cover**
	Jack Lewis	Jeff Weeks

Minecraft™ Mastery: Build Your Own Redstone Contraptions and Mods

About the Authors

Matthew Monk (Preston, UK) is an experienced Minecraft player with a strong interest in the technical aspects of the game as well as making mods for it. He is currently in high school. Matt is also the author of *ComputerCraft: Lua Programming in Minecraft*, and is an active contributor to the Minecraft mod and video community as well as hosting Minecraft servers for his friends.

Simon Monk is a full-time author and loves programming and electronics, ideally in combination. He has written books about the Raspberry Pi, Arduino, and BeagleBone platforms as well as *Hacking Electronics* (McGraw-Hill Education). He spent many years as a professional programmer before turning to writing. You can find a list of his other books here: http://www.simonmonk.org.

You can follow the authors on Twitter using @mcmasterybook.

CONTENTS AT A GLANCE

CONTENTS

ACKNOWLEDGMENTS

We would like to thank Roger Stewart, Patty Mon, Vas Sharma, and the excellent TAB team for doing such a great job in the production of this book.

We also owe a great debt to Mojang for producing such a compelling and flexible game.

Matthew thanks Sue Bradshaw (Mum) for her support during the writing of this book and Simon Monk (Dad) for encouraging him to divert valuable gaming time into the writing of this book.

INTRODUCTION

This book is aimed at the reader who has started playing Minecraft and is ready to get involved in some of the more technical aspects of Minecraft. The book explains how to use redstone to make interesting contraptions. It also explains how to use both the qCraft and ComputerCraft mods.

Taking this even further, the book shows you in easy steps how to start creating your own mods with a little Java programming.

How to Use the Book

The book is designed to gradually progress through levels of difficulty so that you can master these advanced sides of Minecraft. It therefore assumes a basic familiarity with Minecraft. The book is organized into the following chapters:

- **Chapter 1: Introduction** An introduction to Minecraft including basic crafting and forging as well as the options available in setting up a new world and getting started. If you already play Minecraft, then you can probably skip most of this chapter.

- **Chapter 2: Basic Redstone** In this chapter, you will learn the very basics of redstone, such as power and knowing what can be controlled by it, and how to use basic redstone items like levers, buttons, and lamps. You will also make a basic redstone-controlled door.

- **Chapter 3: Redstone Logic Gates** Stepping up a gear from basic redstone, this chapter covers the creation of commonly used logic gates and flip-flops that can be used to build really advanced contraptions.

- **Chapter 4: Advanced Redstone** Applying some of the techniques from the previous chapters, this chapter shows you how to make block swappers, door locks and a functional seven-segment display.

- **Chapter 5: Miscellaneous Redstone** Taking a step back from logic and the more electronic side of Minecraft, this chapter takes a look at some fun and easy-to-use Minecraft items, like minecarts, hoppers, command blocks, and water. It also shows you how to construct a TNT cannon.

- **Chapter 6: Server Hosting and Tools** Being able to share a Minecraft world with friends is one of the great joys of this game. In this chapter the various options for hosting a Minecraft server are explored. This chapter also looks at using the MCEdit tool for offline editing of Minecraft worlds and interfacing with Minecraft on the Raspberry Pi using Python.

- **Chapter 7: qCraft** The qCraft mod allows you to make contraptions and special effects in Minecraft inspired by the bizarre concepts found in quantum mechanics. You can create teleporters, make things appear invisible, and create other interesting effects in this mod.

- **Chapter 8: ComputerCraft** ComputerCraft is a mod that allows you to make projects using computer blocks. You can program the computer in a language called Lua. The computer can also have inputs and outputs that connect to redstone devices. This allows the computer to control mechanical things in your Minecraft world. The mod also includes moving computers called turtles. This chapter shows you how to use ComputerCraft and also how to write simple programs for your computers and turtles in the Lua programming language.

- **Chapter 9: Modding with Forge** In this chapter you will learn how to set up a modding development environment for Minecraft 1.7 using Forge and Eclipse.

- **Chapter 10: Example Mod: Thorium** This chapter takes you step by step through the process of creating a block-based mod including crafting recipes and custom textures.

- **Chapter 11: More Modding Examples** This chapter looks mostly at making item mods, including armor. It also looks at how you can create your own GUI (Graphical User Interface) and link it to right-clicking on a block.

Source Code

All the example code used in the book, including source code for mods as well as Lua programs used in the ComputerCraft chapter, is available as open source software, available either via the book's website (www.minecraftmastery.com) or directly on github at https://github.com/simonmonk/minecraftmastery.

Example Worlds

Following descriptions of how to build complex things in Minecraft is not easy in book format. To help with this, and to overcome the limitations of monochrome books, the book is accompanied by an example world that you can explore. This world includes all the contraptions described in the book, so that you can get up close to them and try them out.

You can download the example world and install it into your saves folder in Minecraft. You will find instructions on how to do this in Chapter 1.

There are also separate example worlds that can be downloaded for the qCraft and ComputerCraft chapters of this book, containing example contraptions for those mods. You can find the instructions for installing them in Chapters 7 and 8.

There are also some videos available showing some of the more complex redstone builds. You can find links to these on the book's website at http://www.minecraftmastery.com.

1

Introduction

Minecraft is an independently produced game that is the brainchild of Markus "Notch" Persson. The game was first released in 2009 and has become hugely popular. The game's appeal is largely due to its creative side, which allows its players to create the most fantastic "contraptions" in a virtual world.

This book is about the creative side of the game, especially using redstone (which is similar to electronics) as well as using and developing your own "mods" to the basic game.

What Is Minecraft?

Minecraft is a first-person game. As a player in Minecraft, you have to survive in a hostile world. This world has night and day and is made up of a vast and randomly generated terrain made up of various types of blocks.

These blocks can be collected and have different properties. You can also transform blocks into more useful items that you can keep in your "inventory." You can use the blocks to create building materials and tools.

However, you are not alone in this world. Your world also contains "mobs." Some of these mobs will come out at night or hide in the shade, and will do their best to kill you (Figure 1-1). These mobs are best avoided by making a home and staying in it at night. As you get better at the game, these spiders, skeletons, and zombies will become less of a threat and you will have crafted yourself tools to fight them with or torches to light an area up and keep them away.

Figure 1-1 *A scary mob*

Other mobs are decidedly useful. These include chickens and pigs, which are sources of food. You need to eat, or you become weak. You can use the wool gathered from sheep (which unfortunately you have to kill) to make things, including a bed to put in your home. A bed might not seem like much of a priority, but Minecraft follows a speeded-up daily cycle. The daytime lasts 10 minutes of real time, as does the night. Since generally at night you probably want to stay indoors, a bed allows you to sleep through the night and speeds up the game play.

You will soon progress past surviving, making yourself a nice secure home, perhaps settling down to farm. If life starts to become a bit pedestrian, you might decide that it's time to do some grand engineering and make some great contraptions just for the hell of it. When doing this, it is a lot easier to dispense with the threat and distraction of mobs, and play the game in Creative mode, where you are indestructible, have unlimited resources, and can fly.

Minecraft Versions

Most people will be using the full version of Minecraft. The game will run on computers using Windows, Mac, or Linux. The game is written in the Java programming language, which is available for almost every type of computer. Java is a fairly resource-hungry language, so despite the fairly simple-looking graphical style of the game, it will play better on a computer with a decent graphics card and plenty of memory. Minecraft is also available for the Xbox and PlayStation 4 and 3 consoles.

A cut-down, simplified version of Minecraft is called Minecraft Pocket Edition. This version is based on an early release of Minecraft ported to the C++ language. It is available on the Android Market and Apple iStore. It lacks some of the sophistication of the full version, including (at the time of writing) redstone, but is actively being developed with new features being added to bring it closer to the full version.

For enthusiasts of the Raspberry Pi single-board computer, there is Minecraft Pi Edition. This has the same basis as the Pocket Edition.

Getting Started

You should start the game playing in Survival mode and understand the basic principles before moving on to more creative endeavors. The key to getting started with Minecraft is to survive your first night. The official guide to doing this can be found here: www.minecraftopia.com/how_to_play_minecraft.

As you progress and need a bit more detailed information, you will find this beginner's guide useful too: http://minecraft.gamepedia.com/Tutorials/Beginner's_guide.

Basic Crafting

Without the ability to make things, Minecraft would be a fairly uninteresting game that mostly involved running away from monsters. The process of making tools and material from things that you pick up in the game (usually after whacking them a bit) is called "crafting."

When you first start the game, the only crafting "recipes" available to you are those that use just four components in the 2×2 grid that you see when you open your inventory by pressing the E key during play (Figure 1-2).

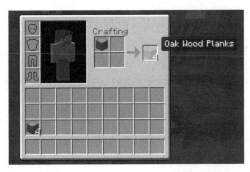

Figure 1-2 *Two-by-two crafting*

Figure 1-3 *Crafting a crafting table*

In this example we have placed a single block of wood (obtained by punching a tree until it breaks) into the top-left corner. This has resulted in four oak wood planks being created from the block of wood. Once the crafted items have been created, you should drag them to the nine slots at the bottom of the inventory. This will make them quickly accessible by typing one of the numbers 1 to 9 corresponding to the slot you want to use.

You can use these blocks to build a shelter, or more importantly to create what should be your first tool, a "crafting table." A crafting table is created by placing four of these planks into the crafting grid (Figure 1-3).

You will normally create a crafting table in your home where you can use it without the danger of attack from mobs. The crafting table gives you a 3×3 grid, which you use in just the same way as the 2×2 grid, but it opens up a lot more recipes.

You will find quite a few of the more common crafting recipes reproduced in this book for your convenience, but the sheer scale of the Minecraft game makes it impractical to include all the crafting recipes. You can find a complete list here: http://minecraft.gamepedia.com/Crafting.

Forging

One of the first things that you will want to make with your crafting table is a "furnace." A furnace will allow you to make even more things, as well as cooking raw meat to reduce the chance of food poisoning (yes, really). You create a forge using a ring of cobblestone. You will have to work a little to get the cobblestone, possibly even doing some actual mining using a pickaxe, which you can make from planks and sticks (sticks are made from planks).

Forging (or cooking) is similar in concept to using a crafting table, because you are transforming some raw material into something else. However, it differs because it requires fuel to burn. For this you can use anything made

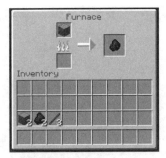

Figure 1-4 *Using a furnace to make charcoal*

of wood, coal that you mine, or charcoal that you make from wood. For example, Figure 1-4 shows how you can make charcoal (useful for making torches) by placing planks in the box below the flames and a block of wood above the flames. Note that the planks placed in the box below the flames have immediately disappeared from Figure 1-4 as they burned.

Throughout this book, you will find pictorial crafting recipes that show what you must place in the 3×3 grid in order to craft some special type of material or tool.

Creative Mode

When you create a new world, you can specify that you wish to play it in Creative mode rather than Survival mode, by clicking the Game Mode button until it toggles to Game Mode Creative (Figure 1-5).

```
Create New World

World Name
Creative
Will be saved in: Creative

Game Mode Creative
Unlimited resources, free flying and
destroy blocks instantly

More World Options...

Create New World          Cancel
```

Figure 1-5 *Setting the Game Mode to Creative*

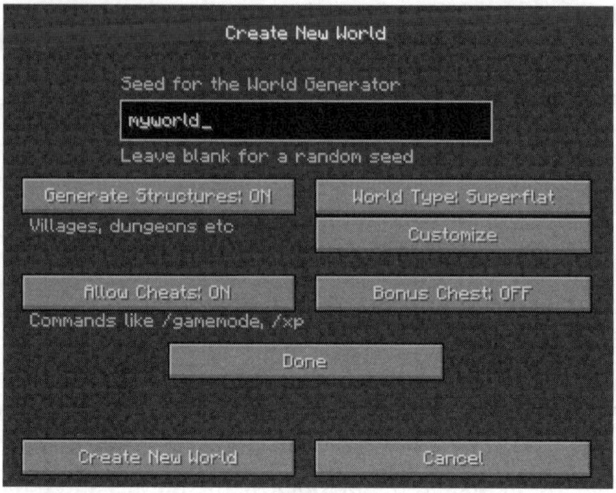

Figure 1-6 *World Options screen*

You can also click the More World Options button at this time to make life even easier (Figure 1-6).

You can optionally specify a "seed" for the world. You can put any text in here and whenever you make a new world with this same seed, you will get exactly the same starting world, in terms of terrain. You could use this feature so that your friends could start with exactly the same world as you.

Another option that you might want to set here is the World Type. Toggling this to Superflat will create a world without mountains and valleys, making it more suitable for building your contraptions on. You can also set the option to Allow Cheats.

Playing in Creative mode will give you the following advantages:

- You will be able to fly.

- You will be invincible.

- You will have an infinite inventory, stocked with every type of item to an unlimited quantity.

Some of the constraints of Survival mode still apply, so you will still have a day/night cycle and there will still be mobs. However, the mobs will not be able to hurt you and you can also turn them off anyway.

Cheats

By selecting the option to allow cheats, you can make the process of creating contraptions even easier. With cheats enabled, you can enter commands by just typing a slash (/) followed by the command.

For example, the following command will set the time to dawn, useful when you want to skip the night.

```
/time set 0
```

There are many commands available and you can see the full list here: http://minecraft.gamepedia.com/Commands.

Redstone

You can make magnificent houses and do all sorts of grand civil engineering projects using basic blocks and tools. However, sooner or later, you will feel the urge to experiment with what in the real world would be mechanical and electrical engineering—that is, making things that move and controlling them electrically.

Redstone is a special type of rock that you can use to make things that are similar in concept to electrical wiring and even electronic components. You can create switches that control torches and move pistons. People have even created entire computers from first principles using redstone.

Using Mods

Many people have developed expansion mods that can be installed into Minecraft. This process is entirely unofficial and not supported by Mojang, the developers of Minecraft. Mods are available to download and range from simple new types of block or modifications to existing items, all the way through to complex mods like ComputerCraft, which we will use in Chapter 8.

Installing a mod is not as simple as just copying some files into the Minecraft installation directory. You have to run a utility program that converts the compiled Java class files back into source code. The source code for the mod is then installed over the top of the original decompiled Mojang code and the files recompiled.

Mods are often available in bundles called *mod packs* that include a whole load of mods from different sources. These normally have an automated installer that takes care of all the class file hacking that is needed.

You can find a list of popular mod packs here: http://minecraft.gamepedia .com/Mods/Mod_packs.

Of these, Feed the Beast and Tekkit are the most popular choices.

Making Mods

In addition to using other people's mods, you can also create your own mods. Chapters 9 to 11 of this book deal with this complex and interesting activity.

You will need to do some Java programming to create mods, but, you do not need to already know Java to be able to follow the examples in this book, which take you step by step through the creation of mods.

Installing the Example World

All the example contraptions described in this book are set up in an example world that you can explore as you read the book. In fact, it will make it much easier to follow the instructions in the book if you download and install this. It will allow you to interact with the contraptions that are built, walk around them to view them from different angles, and even pull them apart to see how they work.

The first step is to download the zip file of the world from: https://github .com/simonmonk/minecraftmastery/tree/master/example_world.

To download the zip file, you need to first click on the link "Minecraft_ Mastery.zip" and then on the page that opens click on the "view raw" link. Save the zip file on your desktop and unzip the file. You then need to copy the entire Minecraft_Mastery folder into your saves folder.

The location of your saves folder will depend on which platform you are running Minecraft on. In both Mac and Windows, these files are hidden by default.

There is a handy trick that you can use to find the saves folder, which works on all platforms, including Windows. The trick is to select Options from the main menu of Minecraft, then click Resource Packs, and then Open

Resource Packs Folder. Then navigate up one level and you will see the Minecraft folder, which also contains the saves folder.

When you have copied the Minecraft Mastery folder into the saves folder, you will find that when you start up Minecraft again, the new world will be there. You can open it and explore the world as you please. If you accidentally damage one of the contraptions that you are looking at, you can always copy the files again.

Summary

In this chapter, we have introduced Minecraft and surveyed the wide range of ways in which you can use Minecraft and even extend it by creating your own mods.

2

Basic Redstone

In this chapter, you will learn the very basics of redstone, such as power and knowing what can be controlled by redstone. If you've played Minecraft before and have a fairly good knowledge of redstone and how to use it, then feel free to skim through this chapter.

Redstone is rather like wires in electronics. It is the thing that allows blocks that control power (buttons, pressure plates, tripwires, and levers) to communicate with blocks that do something with the power, like torches, lamps, and pistons. By combining these blocks and linking them up with redstone, you can create complex contraptions.

The Very Basics

Before we look at some of the blocks related to redstone, let's cover the very basics.

Obtaining Redstone

If you're playing in Creative mode, then redstone dust can be found in the Creative menu under the Redstone tab. On the other hand, if you're playing in Survival mode, then you will need to find redstone ore.

You cannot find redstone ore just anywhere. Redstone ore can be found at Y Level 16 and below, and requires an iron pickaxe or better to be mined.

Redstone ore can drop either four or five piles of redstone dust, but when mined with a Fortune III pickaxe, redstone ore can drop from four to eight piles of dust. An interesting property of redstone ore, unique to the block, is

Coordinates in Minecraft

Minecraft uses a system of coordinates for the positions of things in the Minecraft world. The X axis lies East to West, the Z axis (not the Y axis) is the position North-South on the horizontal plane, and the Y axis is the depth or height (the elevation).

You can find your current position by pressing the F3 key, or CTRL-F3 if you are using a Mac.

that if a player left-clicks or right-clicks on it, or if an entity moves or falls onto redstone ore, then it will emit level 9 light until it receives its next block tick, which is random, but on average is 47 seconds. Because the block changes to a different block, it is very useful for Block Update Detectors (BUDs), which are explained in Chapter 4.

Placing Redstone

Now that you have some redstone dust, you'll want to place it. To do this, simply aim your cursor where you want to place it and right-click, as with any other block. Redstone dust can only be placed on top of a block, but you can't place it on every block. Redstone cannot be placed on transparent blocks, or blocks that are not full. Transparent blocks include any block that a player can see through (even if they can only see through it slightly). Redstone cannot be placed on half slabs placed on the bottom of a block. However, it can be placed on a half slab placed on the top half of a block. When a piece of redstone is placed adjacent to another, they connect up in a straight line. If redstone is placed one block to the side and then one block down, it will connect to redstone on top of the block one block up and to the side. It will appear as a dull red color until you supply it with power.

Let's start by laying out some redstone, as shown in Figure 2-1.

Notice how, as you lay the redstone around a corner, it makes a right-angle bend in the redstone. The redstone is still just a dull red pattern in the sand, so in the next section we will discover how to power it.

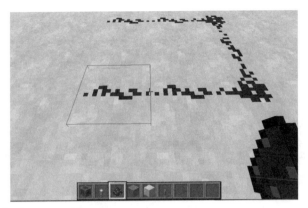

Figure 2-1 *Laying out redstone*

Redstone Power

When redstone dust is supplied with power, it glows red and emits a sparkling animation. The furthest redstone power can go is 15 blocks. It will get progressively dimmer the further away it is from the power source. If you wish to carry a signal or power of more than 15 blocks, either redstone repeaters are required.

As we describe these various sources of redstone power, you might like to try them out in your own world, or in the downloaded example world that we mentioned in Chapter 1.

Levers

Levers are made by putting a stick above a piece of cobblestone in any crafting space. Similar to buttons, they are also found in the Redstone tab when in Creative mode. Levers can be placed anywhere on a block, but only on the same blocks as redstone except for glowstone and ice, which a lever can also not be placed on.

Levers act as a two-way switch, meaning that when the lever is right-clicked, it toggles between an on and off state. Add a lever to the end of the track of redstone that you laid out, so that it looks like Figure 2-2.

Now when you right-click on the lever, it will toggle between on and off, and when it is on, the whole length of the redstone track will light up in bright red.

Figure 2-2   *Powering redstone with a lever*

Buttons

In Creative mode, buttons can be found under the Redstone tab in the Creative menu. In Survival mode, to craft a button, simply follow the recipe shown in Figure 2-3. You can do the same with stone instead of wood to make a stone button.

Buttons have the same rules as redstone for where they can be placed, except buttons can only be placed on the sides of blocks. To activate a button, right-click it. It will play a sound effect, depress into the block, and then pop back out, ready to be pushed again. During all of the time it is pushed in, it will supply redstone power. This length of time depends on whether it's a stone or a wooden button. Stone buttons supply power for one second, whereas wooden buttons supply power for 1.5 seconds. Buttons cannot be activated by mobs unless a skeleton inadvertently shoots a wooden button with an arrow, as wooden buttons can be activated with arrows fired from dispensers, players, or skeletons.

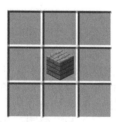

Figure 2-3   *The crafting recipe for a button*

Figure 2-4 *Using a button with redstone*

To replace the lever in our example with a button, we will need to add a block, as the button cannot be on the top or bottom of a block. Figure 2-4 shows this arrangement.

Pressure Plates

As you would expect, pressure plates send out a redstone signal when something goes on top of them. There are currently four different types of pressure plates; two of them are activated only by items, and the other two are activated by players, mobs, and items.

Weighted Pressure Plate (Light)

The crafting recipes for all of the pressure plates are shown in Figure 2-5.

The Light Weighted pressure plate has a Gold texture, because it is made of gold ingots. For roughly every four blocks thrown onto this pressure plate, it will emit a stronger signal by one block. Maximum signal strength (15 blocks) requires you to have 57 items on the pressure plate.

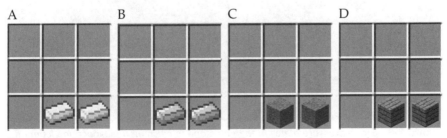

Figure 2-5 *The crafting recipes for all four different types of pressure plate*

Weighted Pressure Plate (Heavy)

You will need iron ingots to make the Heavy Weighted pressure plate, as this one has an iron texture. Every time approximately 42 blocks are thrown onto this pressure plate, the signal strength increases by one. Therefore, to achieve maximum strength with the heavy version, you will need to put 598 blocks, or nine stacks of 22 blocks, on it.

Stone Pressure Plate

Stone pressure plates are made the same way as the weighted pressure plates but with stone (not cobblestone). When a player or a mob stands on them, they emit a redstone signal on all sides (except upward) until that player or mob steps off them again. This makes them extremely useful for doors, or as traps, as they blend in so well (Figure 2-6).

Wooden Pressure Plates

Wooden pressure plates are made in the same way as the others, and you guessed it, they're made with wooden planks instead of stone or gold or iron! This block behaves similarly to its stone counterpart, except it has more ways to be activated. In addition to a mob or a player triggering it, when an

Figure 2-6 *Stone pressure plates hidden on stone to catch an unsuspecting victim off guard*

Figure 2-7   *The Tripwire hook crafting recipe*

item is thrown on it, the Wooden pressure plate also emits a maximum signal strength on all sides (except upward). So it could be thought of as an extremely Light Weighted pressure plate. As with a wooden button, when an arrow hits a wooden pressure plate, that also triggers it.

Tripwires

A *tripwire hook* is made by putting an iron ingot on top of a stick on top of a wooden plank in a crafting table (as shown in Figure 2-7). This will give you two tripwire hooks. As its name suggests, this block acts as a tripwire. To use it, simply put a tripwire hook at both ends of the desired space, and place string from one hook to another (see Figure 2-8 for a visual demonstration). Once almost any entity (except Ender Eyes, thrown Ender Pearls, and thrown potions) crosses the string, a redstone signal is output through the blocks that the hooks are attached to. Tripwire hooks are therefore commonly used in traps, as in low light levels the string can be very hard to see.

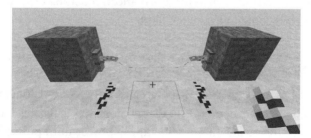

Figure 2-8   *A simple Tripwire setup*

The tripwire will power redstone at either hook, as a pulse whenever the tripwire is crossed. Triggering the tripwire does not break it, and the only way to deactivate tripwires is to break the string with shears (if shears are not used, it will activate the hook) or to find the hook and then destroy either the hook or the block that it's on.

Redstone Blocks

A fairly recent addition to Minecraft is the block of redstone. A block of redstone is crafted by filling the 3×3 area of a crafting space with redstone dust, and like all of the items mentioned in this chapter, it is also found in the Redstone tab. The redstone block acts the same as any other generic block in the game (such as dirt or cobblestone), but it also supplies redstone power on every side of the block. A piston can be used to push or pull it, which can be very useful, and is something you will learn about later on in the book.

You could think of this rather like a large battery that never runs out of power.

The Redstone Tick Rate

Minecraft runs on a tick rate. In reality, most of the time, the smallest measure of time we use is a second. Minecraft uses ticks instead. Every 0.05 second is a Minecraft tick. Confusingly, a redstone tick is every two game ticks, so every 0.1 second. Almost all the time, you will be dealing with one redstone tick or greater, but it is possible to get a half redstone tick using comparators; however, this is currently not used very often. This becomes most relevant when you start to look at more advanced redstone techniques such as "clocks." We will come to these in Chapter 3.

Torches

Redstone torches are very useful and are seen in most redstone builds. This is because of one property unique to the redstone torch: it acts as an inverting power source. This means that it will supply power until power is supplied to it and then it will switch off.

Figure 2-9   *The inverting action of a torch*

Figure 2-9 shows a torch on a block connected to a lever. When the lever powers the redstone between the lever and the torch, the torch is off. When the lever is flipped, the redstone stops being powered but the torch turns on.

Of course, this can be used in a variety of different ways; for example, if you need something to be off at the same time as something else needs to be on, you could use a torch to invert the signal. Figure 2-10 shows a redstone torch being used to "invert" its input. Note that for this to work, the torch must be on the side of a block and the redstone output to the inverter needs to be one block out from the side of the block to which the torch is attached. This is because the torch is occupying the space of the adjacent block.

Another example is that a redstone torch can be used to make a one-block-wide redstone stack (Figure 2-11). This means you can carry the redstone signal vertically in a one-block-wide construction by alternating blocks and torches. Figure 2-11A shows the stack and a tower of "scaffolding" to make it easier to place the alternating blocks and torches. Figure 2-11B shows the stack without the scaffolding.

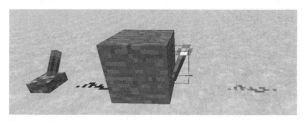

Figure 2-10   *Using a torch as an inverter*

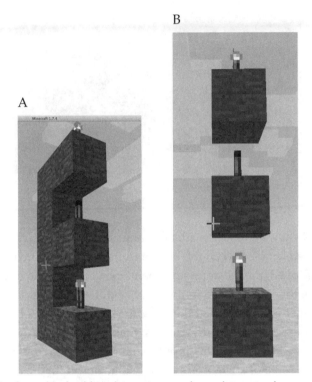

Figure 2-11 *A one-block-wide redstone tower using redstone torches*

The only downside of this is that a redstone torch takes a redstone tick, or 0.1 of a second, to "react." Although this doesn't seem like much, it can soon add up, and it's a good idea to pay attention to this, especially in more complex builds, where the delays may cause problems.

Another, rather unusual, feature of redstone torches is burnout. Burnout occurs when a redstone torch is turned on and off, repeatedly, very quickly. The torch will then remain off until a random tick that could take 30 seconds before it will work again.

In addition to the useful redstone properties of redstone torches, they can also be used as actual torches. Although they only emit low-level light (7), they cannot melt snow or ice. But, because they only emit level 7 light, it is not sufficient to prevent mobs from spawning.

Figure 2-12 *The crafting recipe of a redstone lamp*

Redstone Lamps

A redstone lamp will output light when activated by redstone. That is, if the redstone dust leading up to it is powered, then they will light. This makes them very useful, as they can simply be used as an indicator to make sure that something is working correctly, or they can also be used as the end product of a redstone circuit, like the light switch shown earlier. The crafting recipe is shown in Figure 2-12.

A redstone lamp is not the only way to light something with redstone. A piston can be used to push and pull a block in front of a light source, as shown in Figure 2-13.

This uses a piston (see a later section in this chapter) to move glowstone up and down in front of an aperture, so that it is only visible when the piston is activated. We will meet pistons later in this chapter.

A   B

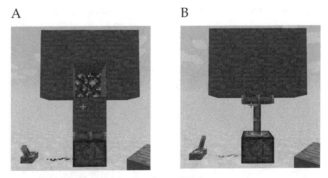

Figure 2-13 *An alternative redstone activated light*

Figure 2-14 *The crafting recipe of a redstone repeater*

Redstone Repeaters

Redstone repeaters are another fundamental component in any redstone contraption, but are fairly heavy on resources (crafting recipe in Figure 2-14). They have many different features that allow them to be used in a variety of different situations. Firstly, redstone repeaters are the easiest way to extend a redstone signal. Remember that redstone signals only travel 15 blocks before they fade out.

As long as there is active redstone next to a repeater, it will boost the signal strength back up to 15 (see Figure 2-15). However, just like redstone torches, they will also delay the signal by one redstone tick.

This is because redstone repeaters also act as delays. By right-clicking on a redstone repeater, you will change the delay from one tick (default) all the way to four redstone ticks of delay, as demonstrated in Figure 2-16.

To change the delay of the repeater, right-click it and you will see that the second peg moves. When it is closest to the main peg, the delay is one redstone tick, and when it is farthest away, it is four ticks.

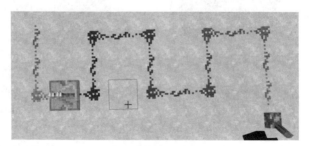

Figure 2-15 *A redstone repeater boosting signal strength*

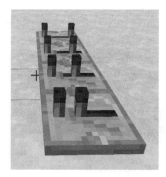

Figure 2-16   *The different delays of a redstone repeater*

Repeaters also act as one-way valves or electronic diodes, meaning it outputs from the front, and the input is in the back. Because of this, redstone can only go one way, and not the other. This can be useful if you need a signal to go through one way, but you don't want it to travel backward. To try this out, build the arrangement shown in Figure 2-17.

You will find that although the left-hand lever turns power to the redstone on and off on both sides of the repeater, when you change the right-hand lever, it only turns on the power as far as the repeater. The redstone to the left of the repeater will not light.

This setup is a good way to experience the effect of delays in redstone ticks. If you push the delay up to its maximum and then flip the left-hand lever, you will see a noticeable delay before the redstone on the right of the repeater catches up with the redstone on the left.

If you power the side of a repeater (A) with another repeater (B), repeater A will be locked in whatever state it's currently in (so on or off) until repeater B is unpowered, and then repeater A will function as it normally would again (see Figure 2-18 for reference).

Notice how the bar appears across repeater A when it is locked.

Figure 2-17   *Using a repeater as a diode*

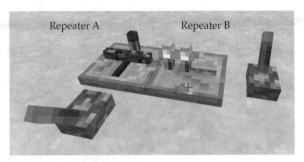

Figure 2-18   *The repeater "locking" effect*

Comparators

Comparators are another fairly recent addition to the game, and require quartz to be built (full crafting recipe in Figure 2-19).

Their main intended use is to compare two inputs and output the strongest signal. Make up the arrangement shown in Figure 2-20 to try out a comparator.

The comparator has two inputs, and in this example, both inputs are connected to levers. With the lever at the front in the off position, the left lever will turn the output of the comparator on and off. However, if the lever at the front of Figure 2-20 is put toward the left (Figure 2-20B), the signal from that lever will be stronger than that of the lever on the left, because it is only one block rather than two away from the comparator. This then turns off the output, as the main input next to the two pegs is no longer stronger than the input connected to the front lever.

A far more common use of comparators is mechanical. This is because they can be used to output different signal strengths depending on, for example, how many blocks are in a container to which they are connected. This might be used to disable a hopper (see later in this chapter).

Figure 2-19   *A comparator's crafting recipe*

A

B

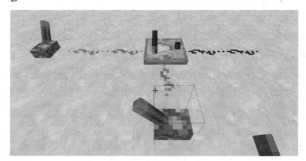

Figure 2-20 *Using a comparator*

Working out how to get the right signal strength out of containers involves this formula:

1 + ((the number of items in the container / maximum stack size for that item) / number of slots in the container) * 14

The answer to this is then truncated, which means you ignore the decimal (if it is 3.00001, then the answer is 3, or if it's 3.99999, then the answer is still 3).

For example, if I put 222 blocks of cobblestone in a dispenser, the comparator will output a signal strength of 6.

1 + ((222 / 64) / 9) * 14 = 6.39583 or 6

Pistons

Pistons are one of the most used redstone components and the crafting recipe is shown in Figure 2-21. They only have one basic purpose, but they are incredibly useful. A normal piston will simply push blocks, and a sticky piston will push and pull blocks. Both types of block move when activated by redstone.

A   B

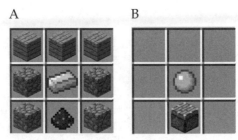

Figure 2-21   *The crafting recipes of a piston and a sticky piston*

Try connecting a piston to a lever by redstone as shown in Figure 2-22. To control which way the piston faces, you will need to move round to the right so that you are facing toward the line of the lever and redstone. The piston will be placed with the moving part towards you.

When you flip the lever, the piston will push the block away. However, when you flip the lever back, the piston will retract, but the block will not be pulled back. Try swapping the piston for a sticky piston. This time, you will find that the block stays attached to the piston.

Pistons can push up to 12 blocks, not including bedrock, obsidian, furnaces, noteblocks, jukeboxes, chests, and some others. Also, sticky pistons will not hold sand or gravel blocks from falling as they normally would. If a sticky piston receives a monostable pulse (a one-tick-long pulse described in greater depth later on in the book), it will "spit" the block out in front of it. If this block is "spat" over a torch, then you can create a device that, when

A

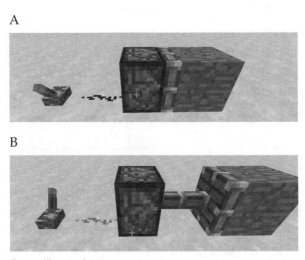

B

Figure 2-22   *Controlling a piston*

activated by a button, will toggle between an on and off state. This is called a "T-Flip-Flop" and we will find out more about it in Chapter 3.

In addition to being a core part of a redstone circuit, a piston can also act as an actuator. A very common use of pistons is to make doors. These doors can simply be one block in the corner of a room to a 4×4 door.

Making a Door

Let's put together what we have learned so far to make a door using a lever, some redstone, and a piston. Start by making the arrangement shown in Figure 2-23.

Redstone can only be placed on the top of a block, so we will need a block to the left of the bottom piston to allow the redstone signal to reach the upper piston. We have used sticky pistons, with a block attached to each.

When you flip the lever, both pistons should activate, pushing or pulling the blocks that will be the door.

We now want to put some blocks in front of the door mechanism that would be the wall of our shelter. The problem is that our redstone is on the surface, so if we placed a block of stone on top of it, it would float above

A

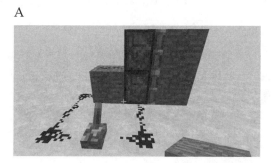

B

Figure 2-23 *A two-piston door*

the redstone. We need to bury the redstone. To do this, dig out the block and place redstone in the hole (Figure 2-24A). You can now build the wall where we want, so the finished door looks like Figure 2-24B.

If we wanted to go a step further and hide all the redstone, then we would need to dig down two layers, so that we could then put a floor block back into the hole (see Figure 2-24C).

There is more on making piston doors, complete with locks, in Chapter 4.

A

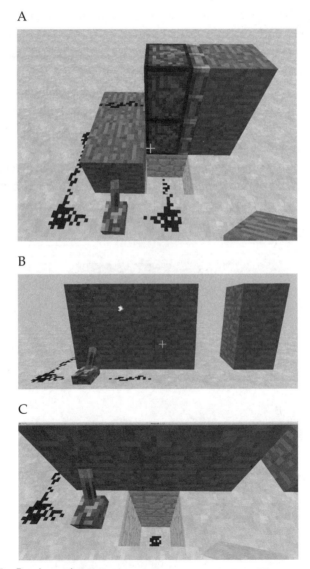

B

C

Figure 2-24 *Burying redstone*

Hoppers, Dispensers, and Droppers

Hoppers, dispensers, and droppers offer various ways of releasing things into the world, either automatically or controlled by redstone.

Hoppers

A hopper will by default allow things above it to drop through below it, automatically. When redstone power is applied to it, this will stop the dropping, "disabling" the hopper.

Dispensers

Dispensers are a little like hoppers, but rather than allowing things to fall through them in a controlled manner, they have to be stocked with an inventory. They will then dispense things from their stock when stimulated by a redstone signal.

For example, Figure 2-25 shows a button controlling power to a dispenser that has been stocked with arrows.

To fill the dispenser with arrows, right-click on the dispenser and place arrows (or anything else you want to drop) into the 3×3 grid at the top.

If you want to try this out in the example world, then stand back a little as you click the button, so that you can see the arrows come flying out. This could of course be combined with a pressure plate or tripwire to make a trap worthy of an Indiana Jones movie.

Figure 2-25   *Firing arrows from a dispenser*

Droppers

Droppers are similar to dispensers, but instead of releasing blocks fully formed, they "drop" blocks and tools just as if a player had dropped something. This can be used as a way of gifting things to other players.

Summary

You can find the crafting recipes for droppers and all the other types of block in Minecraft at http://minecraft.gamepedia.com/ and search for the block or tool that you are interested in making.

In this chapter so far, we've covered the very basics of redstone. We know the different ways of powering redstone, the different parts of redstone circuits, and also things that can be activated by redstone.

Now that we know exactly how the basic components of redstone work, we can start to look at the more advanced parts, as well as building our first contraptions. To do this, we are first going to look at logic gates, which are also found in most electronic devices and are the building blocks of real-world computers.

3

Redstone Logic Gates

In this chapter, you will learn all about logic gates. Almost every complicated redstone circuit you will ever build will have at least one of these.
You will also learn about T-Flip-Flops, and "clocks," which all link in close-
ly with logic gates.

Logic gates are also found in electronics, so you may have already heard
of some of these. We will be covering AND, OR, and XOR gates, as well as
inverters, RS NOR latches, and flip-flops of various types.

In Minecraft, logic gates do not come readily assembled; you make them
using various redstone blocks.

Basic Logic Gates

Because logic gates are very commonly used, people have developed inge-
nious and compact ways of making them. All of the gates we'll be covering
rely on redstone torches, but there are alternative methods using pistons and
redstone blocks. We'll look at how to build them, how they work, and how
they can be used.

Inverters

An inverter is a type of logic gate that has a single input and a single output.
The output is always the opposite of the input. So, if its input is on, its output
will be off and vice versa.

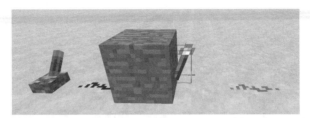

Figure 3-1   *A redstone torch inverting a signal*

Inverting redstone signals in Minecraft is very easy. Simply run the line of redstone that you want to invert into a redstone torch, and use the output of that torch as your new line of redstone (demonstrated in Figure 3-1).

This works because (as we saw in Chapter 2) torches are natural inverters. That is, they will be lit unless they are powered by redstone at their input.

AND Gates

AND gates can have two or more (usually two) inputs and one output. For the output to be on, all of the inputs must be on at the same time. This is very easy to do in Minecraft using torches, and you can see the design of a two-input AND Gate in Figure 3-2. The inputs to the AND gate are levers on the left of the blocks and the "output" torch is on the right. If the torches are on at the input, then the levers are off because the torches are inverters. You would normally also have redstone signals leading to and from the inputs and outputs.

A   B   C

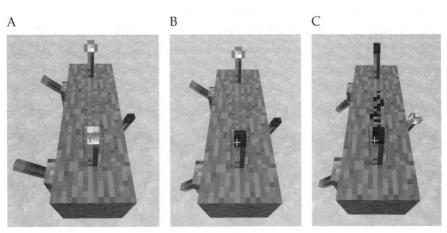

Figure 3-2   *A two-input AND gate*

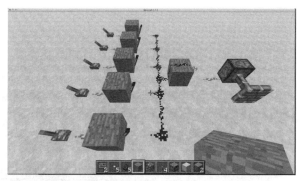

Figure 3-3 *A five-input AND gate*

Figure 3-3 shows a five-input AND gate, with levers connected to all the inputs and a piston connected to the output.

It is fairly easy to see how both of these designs work. In Figure 3-2, the two torches on the top power the redstone on the middle block. This disables the torch on that block until both of the other torches are turned off. The circuit in Figure 3-3 works in the same way. All of the input torches power the same line of redstone. That line of redstone is powering the end torch, and when all of the torches are turned off, or the levers turned on, the output will also be on.

A rather crude example of using AND gates is a pass code lock (shown in Figure 3-4). When the correct levers are flipped, the door opens. This is because two of the levers are connected to the AND gate and the other two

De Morgan's Law

The design of Figure 3-3 makes use of a law of logic called De Morgan's Law. This is best explained as:

> Inverting the result of two inputs "anded" together gives the same result as inverting both the inputs and "oring" their outputs together.

In Figure 3-3, what we actually have is all of the inputs being inverted by a torch, "ored" together, and then the output of the "or" gate is then itself inverted, giving the overall result of "anding" all the inputs. This is a useful trick to know.

A

B

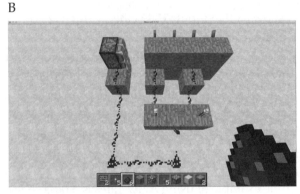

Figure 3-4 *A pass code lock using an AND gate*

levers are dummy levers not connected to anything. However, this lock is very insecure because, if all of the levers are set, the door will still open anyway. This is something you will learn how to fix later on in the chapter.

Or Gates

OR gates are similar to AND gates in that they both have two or more inputs and one output. When one or more of the inputs are on, then the output will also be on. So this means as long as any of the inputs are on, the output will also be on. OR gates are the simplest gates to make in Minecraft. You do not need any torches; you just connect together a number of input signals with redstone dust. Figure 3-5 shows a five-input OR gate.

Try making it, or try it out in the book's example world to confirm how it behaves when you change the inputs.

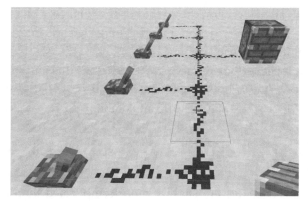

Figure 3-5 *A five-input OR gate*

One example of a use of an OR gate is if you want to switch something on from multiple locations. Here, you can see a very simple demonstration as a light switch (Figure 3-6).

A

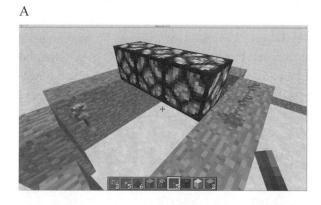

B

Figure 3-6 *A light switch with multiple inputs*

Figure 3-6A shows the building without the final roof block in place, so that you can see the construction. Figure 3-6B shows the interior placement of the light switches by each door.

Each of the levers is wired together with redstone that joins together at the redstone lamps in the ceiling. Just as in real life, having a light switch perhaps next to each door saves having to fumble about in the dark for the light switch.

XOR Gates

XOR (exclusive or) gates are slightly more complicated than the previous two, but are still fairly compact. They can only have two inputs, and one output. XOR means either or, so the output will only be on when one of the inputs is on. If both are on, it will not output a signal. You can see how to build an XOR gate in Figure 3-7. This one has two levers to set the inputs, and the output controls a piston.

To build an XOR gate with this pattern requires six blocks and seven torches!

It uses an AND gate, as well as an OR gate. So when one input is on, the output will also be on, but because of the AND gate, when both inputs are on, the AND gate outputs and disables the output of the XOR gate. When you need an XOR gate, you can just copy this design. The easiest way to understand it is to build it yourself and play around with it.

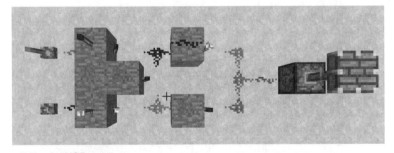

Figure 3-7   *An XOR gate*

Logic Gate Example

As an example of just how useful logic gates can be, let's build a door lock that requires a certain combination of four levers to be in the right positions. Figure 3-8 shows the completed contraption from above so that you can see how it works. This is an improvement on the example from Figure 3-4, because this lock does actually only unlock when all four switches are in the correct position.

The logic works like this. Two of the four levers have a redstone torch after them, which inverts their signals. The signals from all four levers (two inverted, two not inverted) are or'ed together by the redstone dust behind them. This line of redstone behind the blocks will be powered if any of the levers (inverted or not) are powered. That is, it will be powered for any position of the levers except the position where the two levers with inverted outputs (A and C) are down and the other two levers are up. This is the combination for the door being unlocked, so we need to invert this or'ed signal and use it to control the piston. This is another example of making use of De Morgan's Law (see the sidebar earlier in the chapter).

Build the contraption or go and visit it in the example world (see Chapter 1) and try the different lever combinations to see how the redstone dust changes from being powered to unpowered.

Figure 3-8   *A four-lever combination lock*

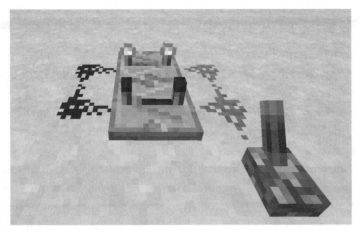

Figure 3-9 *A simple clock*

Clocks

By connecting two repeaters in a loop together, they can be made to oscillate, with the redstone on each side of the repeater powering up in turn (Figure 3-9).

There are a couple of different types of redstone clocks; the most common one uses two repeaters and four redstone. To start the clock, place a lever anywhere next to the redstone and flick it on and off very quickly. It will start more easily if you set the delays on both repeaters to maximum. Once the oscillations start, you can destroy the lever.

If you have both of the repeaters on minimum delay, it can be difficult to start without a monostable circuit (a one-tick pulse covered in the next chapter). If you don't flick the lever fast enough, the redstone will get stuck on until you break some of the redstone. This is called a *feedback loop*. To extend this type of clock, simply add more repeaters in both directions (see Figure 3-10).

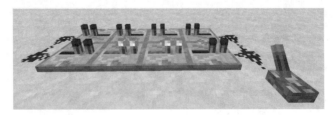

Figure 3-10 *Adding more repeaters to a clock*

Redstone clocks are used in a lot of bigger builds, as they can be used to trigger certain parts of a contraption at certain times. Timing is very important, and using clocks, you can accurately determine when certain things should happen.

The basic clock isn't particularly useful because it's not easy to turn on or off. This is where you get variations. The simplest to make that can easily be toggled works on a similar principle to the first one, but also uses a torch. You can see an example of this in Figure 3-11. By powering the torch, you can stop and start the clock far more easily. To extend the time it takes the clock to cycle, make the circuit longer by adding more repeaters. I would recommend using this clock all of the time unless you need a one-tick pulse clock that will not work with this circuit because the redstone torch cannot switch on and off fast enough.

Of course, there is no reason why you cannot attach something to the redstone to make it flash as the clock oscillates. Figure 3-11 also shows how you can use the clock to make some torches on a block flash. You could bury the clock so that only the flashing light is visible, or have the clock control a piston as you will find in the book's example world. You will find it just behind Figure 3-12.

If you want to see how this works, find it in the example world and carefully excavate around it to uncover the hidden redstone.

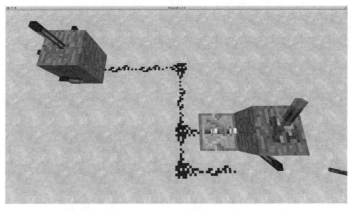

Figure 3-11 *A redstone clock that can easily be turned on and off*

Figure 3-12   *A hidden clocked piston with control lever*

RS NOR Latches

RS NOR latches are used quite often in redstone builds. RS NOR is short for Reset Set Not OR. This seems complicated, but the end result is fairly basic. The design itself is also very basic. It is simply two redstone torches feeding into each other (Figure 3-13). The design is symmetrical; each block has a redstone torch on the front or back and a button on the side.

Figure 3-13   *The most basic RS NOR latch*

When the button on the block of the torch that is on is clicked, the other torch will turn on and that line of redstone will be powered. If you click that button again, nothing will happen. To get it back to its original state, you will have to click the other button. This means it acts as a resettable latch, using two inverted OR gates, hence its shortened name, RS NOR latch.

RS NOR latches come in many different shapes and sizes, and most of them are considerably more compact than the first one. A good example of a one-block-wide version is shown in Figure 3-14. It works on exactly the same principle, one redstone torch powering another.

A common use of RS NOR latches is, in mapmaking, if you want someone to be able to do something once, or until you want them to be able to do it again. If you only give them access to one button, after they've clicked it once, it becomes inactive.

A

B

Figure 3-14 *A one-block-wide RS NOR latch*

Figure 3-15   *A monostable design*

Monostables

The clocks we have been looking at will keep oscillating continuously, and their counterparts in the world of electronics are called *oscillators* or sometime astables (no stable state). A related logical building block is the "monostable." Unlike a clock, a monostable will just send a pulse when triggered and that pulse will always be of the same duration.

A push button is a kind of monostable, but sometimes you want to control the "on time" to something other than the time the button stays depressed. You may also want to trigger the monostable from a redstone output rather than the click of a button.

Figure 3-15 shows a common design for a monostable.

In this case, the input is a button and the output controls a piston. You could actually consider the torch on the right-hand block as being the output. The basic principle is that when the button is clicked, or some other redstone signal powers the input, a signal goes to both the right-hand block and the repeater. This immediately powers the output, which will stay powered until the repeater finishes its delay, at which point, it will cancel the pulse. Changing the delay on the repeater will change the duration of the pulse that the monostable generates.

T-Flip-Flops

T-Flip-Flops ("T" for "Toggle") are found in a lot of redstone builds, particularly small ones, like a simple trapdoor or a light switch. They have a similar function to a lever, but are often better. If you've used levers enough, you will know that once the lever is turned on from one side of, say, a door, then the door cannot be closed if the original lever is not returned to the off position. A button, however, would instead open the door for a short period of time before it closes behind itself. Using a T-Flip-Flop, when the button is clicked, the T-Flip-Flop will change state, opening the door, and when a button attached to the same line of redstone on the other side of the door is clicked, it will change state again, closing the door.

The T-Flip-Flop is very useful, and based on a fairly simple concept, so let's look at how to build some.

This first example of a T-Flip-Flop uses a repeater, two comparators, and a sticky piston to which redstone is attached (Figure 3-16).

To be perfectly honest, I do not have a clear grasp of how this works. There is feedback between the two comparators making a basic flip-flop mechanism, but essentially the design (for this author at least) is treated as a black box that simply works. If you just want piston movement as an output, then you could attach other blocks directly to the redstone, rather than have the separate piston on the left of Figure 3-16.

A second example of a T-Flip-Flop (Figure 3-17) is a compact design that has been around for a long time and uses two opposing pistons with a block in between them.

A redstone torch must be placed under the right-hand piston and the output (underneath the button) becomes powered when the block is pushed over the torch.

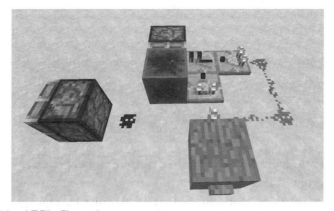

Figure 3-16 *A T-Flip-Flop using comparators*

A

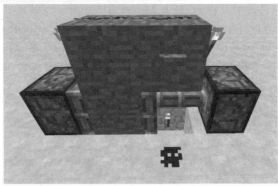

B

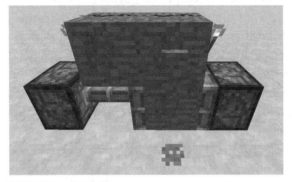

Figure 3-17 *A T-Flip-Flop using pistons*

LogicTheory

There are many parallels between the world of electronics and the world of redstone. Logic gates are the building blocks of computers, so for the curious, here is a quick primer on binary logic.

Binary

The thing about logic gates is that their outputs are either on or off; there is no middle ground. They cannot be "a bit on" or "a bit off"; it's all or nothing. Another way of saying that is to say that the output of a logic gate can either be 0 (off) or 1 (on). A single digit that can only be a 1 or a 0 is called a binary digit. The term "binary digit" is much better known as "bit."

One bit on its own can sometimes be useful, but generally becomes a whole lot more useful when it comes as a collection of bits. A "byte" is a collection of eight bits. Each of those bits can be either 0 or 1. So, a byte might be written in binary as 00101011.

We are used to using the decimal system, where as we look at a long number, going from right to left, the first digit is the units, the next digit to the left is the tens, the next the hundreds, and so on. A bit does not have ten possible values; it only has two (1 or 0) so, when reading a binary number from right to left, the first digit is the "ones," the next to the left the "twos," then next the "fours," and so on.

Table 3-1 shows the result in counting from 0 to 7 in binary.

As you can see, 3 bits allow us to represent the decimal numbers between 0 and 7 (8 different numbers). If we had a whole byte of 8 bits to use, we could count from 0 to 255.

Logic Diagrams

Electronics has a set of special symbols for logic gates. Figure 3-18 shows theses symbols.

An "inverter" is also called a "not gate" for obvious reasons. NAND and NOR gates are AND and OR gates (respectively) with an inverted output. They behave the same as an AND or OR gate with an inverter on the output.

Binary	Decimal
0	0
1	1
10	2
11	3
100	4
101	5
110	6
111	7

Table 3-1 *Counting in Binary*

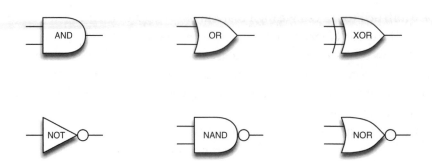

Figure 3-18 *Logic gate symbols*

These symbols become useful when you start to combine them into complex logic circuits. For example, you can use logic gates to add binary numbers together. For a single-bit adder this is really easy. Figure 3-19 shows the two gates that you need.

When thinking about how a logic gate works, or how a number of logic gates work when connected together, it is often useful to use something called a truth table. Table 3-2 shows the truth table for the single-bit adder.

To draw a truth table, you have a column for every input and output, and put a row in for every possible combination of inputs. Looking at the table, you can see that if A and B are both 0, then the sum is 0. If one of them is "1" then the sum will be 1; however, if both A and B are 1, then the Sum digit itself will be 0, but we will want to carry a 1 to the next number position. In binary, 1 + 1 is 0 carry 1 (or decimal 2).

This does mean that if we want to add more than one bit at a time, we will need to have the next adding stage have three inputs (A, B, and carry in).

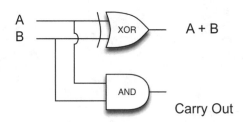

Figure 3-19 *A single-bit adder*

A	B	Sum (A+B)	Carry Out
0	0	0	0
0	1	1	0
1	0	1	0
1	1	0	1

Table 3-2　*The Truth Table for a Single-Bit Adder*

This is a bit more complicated to make, as we have to add together three digits rather than two. Figure 3-20 shows an adder stage that takes a carry-in digit.

If we had eight of these stages, then we could use it to add two bytes together. Every computer's CPU will have a hardware "adder" made up of logic gates in a very similar way to this. A 32-bit processor will be able to process 32-bit numbers in one go, and a 64-bit machine will have 64 stages as in Figure 3-20 to add 64 bits at a time.

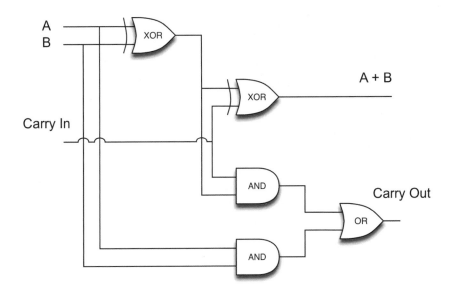

Figure 3-20　*A single-bit adder with carry-in bit*

Summary

We now know all about logic gates and basic redstone, so we are prepared to get to the good stuff, advanced redstone.

We will be combining all of the knowledge we have picked up so far into building complex contraptions, such as making memory in Minecraft with pistons, secret doors, and other advanced contraptions.

4

Advanced Redstone

In this chapter, you will learn some of the more advanced uses of redstone that are generally more resource-intensive but usually far more interesting.

Some of the things you will learn to build include block swappers, concealed doors, and a functional seven-segment display. Don't be put off if you're new to the game and some of the concepts (especially the piston tape) seem confusing at first!

The best way to learn Minecraft is to play. So, try out the examples here and if you get stuck, you can try them out on the downloadable world available at the book's website: http://www.minecraftmastery.com. You will also find links to videos on the book's website showing how some of the more complex contraptions are put together.

Block Updater Detector Switches

Block Updater Detectors, or BUDs, rely on what technically is a bug in Minecraft. However, because BUDs are useful and crucial in many builds, the developers of Minecraft have decided to leave them in. Basically, whenever a block is updated, either by placing something on it, removing it, or changing it to another block (for example, from a closed fence gate to an open one), it triggers a pulse of redstone power.

Figure 4-1 shows a simple and common BUD design.

The piston on the left of Figure 4-1 is a normal piston, but the piston on the right needs to be "sticky." You will probably find out how unstable this design is when you start to build it. As soon as you place the redstone dust,

Figure 4-1   *A simple BUD switch*

you will find that the right-hand piston starts going mad. The secret is not to add the dust until the rest of the contraption has been built.

Once you have made the arrangement shown in Figure 4-1 or visited it in the example world available from the downloads section of this book (www .minecraftmastery.com), try adding a block of stone onto the front of the right-hand piston. You will see that it activates and moves the redstone block, powering the torch and trail of redstone dust to the left of the torch. This is the output of the contraption, and you will notice that it resets itself after a short pulse of power.

Making Piston Doors

Piston doors come in all shapes and sizes, from a 1×1 hole in the ground, to huge 30×11 hangar doors. Because a sticky piston's main function is to push and pull blocks, it is ideal to make hidden doors with them as an alternative to wooden and iron doors. Let's start off with a simple 2×2 sliding door.

Building a 2×2 Door

We see examples of doors like this in the real world every day. This door will automatically swing open as you approach it and step on the pressure plate in front of it.

Although your typical sliding store door is usually made of glass, this simple design will not be made with see-through blocks; otherwise, you would be able to see the sticky pistons, which aren't the most aesthetically pleasing blocks.

Figure 4-2 *Placing the sticky pistons and door blocks*

This door is very simple to make and only requires two pressure plates, four sticky pistons, two redstone torches, some redstone dust, two half slabs, and some blocks.

First of all, place four blocks down that will be the door, and then place your sticky pistons one block away from them (see Figure 4-2).

Then, a couple of blocks in front of the four blocks that will form the door, place your pressure plate. Next, place a block next to both of the top pistons, and then a torch below it. All of the sticky pistons should now be extended (as in Figure 4-3).

The block and torch arrangement will ensure that both the upper and lower pistons will move together when a signal is sent from the pressure plate.

Finally, hook the pressure plate up to the torches (under the surface for the blocks closest to the torches). We have left as much of the redstone wiring as possible on the ground, to make it easier to see what is going on. The finished door is shown in Figure 4-4.

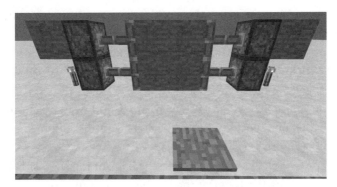

Figure 4-3 *Adding the pressure plate and the door mechanism*

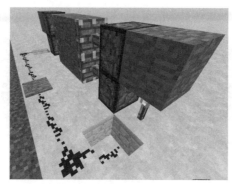

Figure 4-4   *The completed 2×2 sliding door*

And that's it! If you want it to open from both sides, simply mirror what you did with the pressure plates and redstone dust on the other side. In addition to using pressure plates, you could also use a button, and if you wanted to, implement a T-Flip-Flop (see Chapter 3) into the design in place of the button so that the door stays open until the button is clicked again. Go ahead, try it yourself!

Building a 2×2 Flush Door

Our last 2×2 door was all well and good, but it was obvious that there was a door there. If you want a hidden entrance/exit, what better way to disguise it than to make it flush (level or even) with a wall? To do this, we are still going to have the same basic sliding mechanism using pistons as in the previous example, but we are also going to have pistons behind the door that pull it back before pulling the two halves of the door apart. In this way, the door can be surrounded by blocks to make it completely secret.

This contraption is considerably more expensive then our last one, requiring 12 sticky pistons, three repeaters, three redstone torches, a redstone block, and a stack of redstone dust.

To start off, place a 2×2 area of sticky pistons, followed by another 2×2 set of sticky pistons one block away on the right and the left, facing the center sticky pistons (Figure 4-5).

Next, place two blocks diagonally on the back of the right side 2×2 piston sets, with a redstone torch under the top block, and redstone dust on top of

Figure 4-5 *Placing the 12 pistons*

the other block as shown in Figure 4-6. All four of those pistons should now be extended on that side. Repeat this on the left-hand set of the pistons as well. The diagonal blocks on the left-hand side should have the lower block toward the front of the contraption.

Now, place blocks across the top of the middle pistons, and then staircase up one block on the right-hand side, place another block at that level, and then staircase downward for three blocks as shown in Figure 4-7. Note that "staircasing" in Minecraft means creating a diagonal zig-zag staircase type structure from blocks.

Note that it does not matter on which side you do this staircase, it's up to you, but if your staircase goes directly over one of the pieces of redstone for the side door, you will have to swap that torch to another side. Place redstone on all of those blocks you have just added except the last one at ground level on the right, where you need to place a torch on the side of it underneath the next staircase block, shown in Figure 4-7.

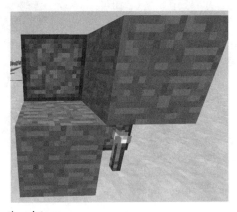

Figure 4-6 *Linking the pistons*

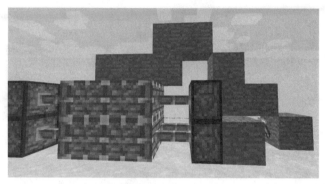

Figure 4-7 *Connecting the middle pistons*

Now we're almost done! It is actually easier to wire up the pressure switch from the back of the contraption and then add a second pressure plate at the front that connects in. So, moving around to the back of the contraption, wire up a pressure pad and repeaters as shown in Figure 4-8.

Change the delay on the repeater to be at least two ticks. This delay is needed so that the door can be pulled back before the side pistons swing into action and open the door.

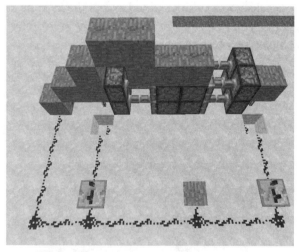

Figure 4-8 *Wiring the back of the door*

Figure 4-9 *Adding a button at the front*

To put a pressure plate round at the front, simply place the plate and run redstone dust around to the back, adding another repeater (as the track is pretty long). This is shown in Figure 4-9.

In the example world, you will find two examples of this contraption, one built into a wall to show how the concealed element of the door works and one exposed, so that you can see more easily how everything works.

This is where the optional parts start to come in. As it stands, there is just enough time for you to step on the pressure plate and then rush through the door before it closes. You could connect the point at which you connected the front pressure plate to a T-Flip-Flop, which will allow you to toggle the door open and shut. We've now made our 2×2 flush piston door!

Block Swappers

As the name suggests, a block swapper will usually swap two different types of blocks on the same position. This is good if you want a hidden crafting bench or hidden glowstone, and so on. We'll be looking at two designs, a very compact wall-embedded design, and a flush one.

The Very Compact Block Swapper

This is a very simple build that takes advantage of BUD switches. You can see how to build it in Figure 4-10 (both pistons are sticky). In the default position,

A   B

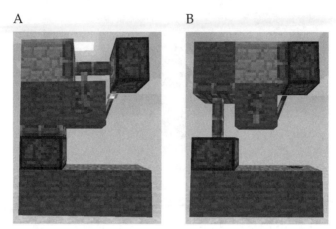

Figure 4-10 *The very compact block swapper*

the torch is powering the top sticky piston and that's it. However, when it is switched on, the top piston will retract, and the lever, which is diagonally powering the bottom sticky piston, creates a BUD, which is updated by the redstone dust below the redstone torch. That process happens in reverse and we get back to our original state. Figures 4-10A and B show the top-left block of the contraption being swapped when you flip the lever.

Note that if you flip the lever really quickly, then it can result in one of the swapping blocks being spat-out rather than moved, which will break the mechanism.

The Flush Block Swapper

The very compact block swapper is fine in many situations; however, it cannot be built so that only the "swapped" block is visible. An example of using a "flush" block swapper might be to create under-floor lighting. When the button is clicked, one of the floor blocks is swapped for a redstone lamp block.

To build such a contraption, we are going to use three pistons. Two push the block to be swapped left to right before the third piston pushes the selected block up into place on the floor. Figure 4-11A and Figure 4-11B show the block swapper in the two positions.

A

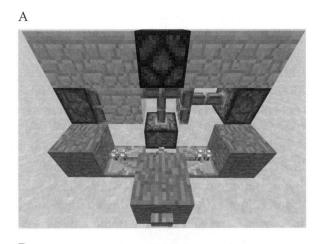

B

Figure 4-11   *A flush block swapper*

This build is slightly larger, but still relatively compact for what it does. You can see how to build it in Figure 4-11. The only thing that may not be obvious from the figure is that there is a torch on the back of the front block with the button on the front. Building the thing can be a little tricky, as the pistons tend to activate when you need them to be retracted to attach the blocks to be swapped. Temporarily removing the redstone dust on top of the side blocks next to the horizontal blocks will make them contract so that you can attach blocks to the business end of the pistons.

Once again, all of the pistons are sticky pistons and the design is symmetrical apart from the block to be swapped. Let's look at how the upward-facing design works. In its normal state, the middle block is simply being pushed upward. When the button is clicked, the middle torch unpowers. First, the middle sticky piston is retracted. Then, as both of the other torches turn off, the one without a block will retract as the other piston extends with its block. The middle sticky piston then fires again, which results in the other block being in the same place as the previous block was. Exactly the same thing happens next time the button is clicked, except the two side pistons do the reverse. If you want to make this block swapper faster, you can connect one of the outputs from the side torches to the middle piston with only one redstone dust, a repeater, and a block at the front of the design.

Piston-Tape Memory

As we learned earlier, logic uses binary, as does memory. Everything is stored in either a one or a zero. We can replicate this in Minecraft, but it calls for quite a large structure. In this case, the memory is going to take the form of a drum (or tape). This drum will be seven blocks wide and information is encoded on the blocks of the drum by using a mixture of normal blocks that transmit redstone power and glass blocks that do not (1's and 0's).

But first, let's build the basic mechanism to rotate the drum memory. Later on we will show how this can be attached to a seven-segment display. Be warned; this is not an easy contraption to build. At least it is not at all easy to describe how to build it, so you will definitely need to look at it in the example world for the book, or follow the video that we have made to accompany the book that shows the build block by block. The great thing about using a video is that you can keep stopping and starting it as you work on building the contraption for yourself. Both the video and the examples for this can be found on the book's website at http://www.minecraftmastery.com.

This is quite a large contraption and would be difficult to build in a fairly new survival world because it requires a lot of pistons and repeaters, so you are almost certainly going to be doing this in Creative mode.

To start off, you will want to make a four-up and four-across ring of blocks. Then, place pistons so that you have two pistons facing horizontally, left and right in diagonally opposite corners, and two facing vertically, up and down, in the other two diagonally opposite corners, as in Figure 4-12.

Figure 4-12   *The first ring of blocks with pistons in place*

You will now need to add another six of these segments next to each other so that the pistons are in rows touching each other. We have laid all the pistons, so now we just need to hook up the redstone. The purpose of this is to push the blocks around in rotations, so the diagonal pistons need to be synchronized. Let's start with the ground-level, horizontal facing, pistons. Place repeaters, with redstone behind them, behind all of those pistons, and for its diagonal counterparts, place blocks on top of the pistons with a line of redstone running across it. For the other sets of pistons, start by placing repeaters facing into all of the pistons in the ground, with a redstone line behind it. For our final corner, place blocks on the same level, and adjacent, to the pistons, with redstone on top of those blocks. Figure 4-13 shows this build in progress, with some of the repeaters in place.

Figure 4-13   *Adding the other six tapes*

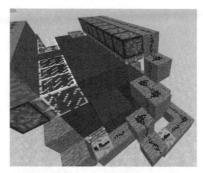

Figure 4-14 *"Staircasing" the connections to the top pistons*

Staircasing the redstone down is a little messy because otherwise it will interfere with another redstone signal that we will get to later. For now, just copy the way it is staircased in Figure 4-14. Note that you can see that some of the blocks of the drum are glass. Replacing ordinary blocks with glass blocks will be how we "program" the drum. For now, you can make all the blocks of the drum normal blocks.

This bit is quite hard to navigate, so be prepared to break and replace some blocks! Inside the ring, place one thick layer of blocks, and then place repeaters pointing at the blocks that get pushed by the upward-facing ground-level pistons. Place a redstone torch at the side with the staircase for the horizontal pistons, and then place redstone dust on top of the rest of the blocks inside there. On the other side of where your repeaters are facing will be our output, but we'll get on to that later when we look at connecting a display to the piston tape memory. Figure 4-15 shows this arrangement.

Figure 4-15 *Adding repeaters inside the drum*

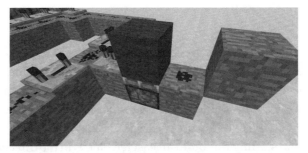

Figure 4-16 *One way to wire the input for the design*

Although it may look as if the repeaters and blocks might get in the way of the drum turning, they do not.

All we need to do to finish it off is have a way of activating the pistons. Just make a simple monostable circuit with a button as its input, and split the output so that the vertical facing piston set has full delay, while the horizontal set has one tick (see Figure 4-16 for an example, where there is a sticky piston underneath the yellow block). Because there are quite a lot of long lines of redstone, make sure that the signal is strong enough to reach where it needs to by using repeaters every 15 blocks of redstone.

Now all you need to do is click the button to test it! We have just made some piston tape/memory, but it's pretty useless by itself. Next we're going to look at how to hook it up to a display. The final contraption is shown in Figure 4-17.

Figure 4-17 *The piston tape contraption*

Displays

This part can get very hard, especially if you're trying to make it compact. You will need to spend a lengthy amount of time routing lines of redstone dust to connect everything up without the lines interfering with each other.

The seven-segment display is going to be attached to the piston tape memory that you made in the previous section.

I recommend starting this build at least ten blocks in the direction of where the repeaters were placed as outputs to the piston tape. Build a nine-wide, eleven-high, one-block-thick rectangle. Now you need to etch the seven-segment display, one block back, with each segment being three blocks long (see Figure 4-18 for reference).

Next, we need to put the sticky pistons in. Put them directly behind each block, so if all of them were activated, it would look like one flush wall again. Hooking up the redstone is a bit more complicated.

The top, horizontal segment of the display is the easiest to make. A single torch will power the redstone linking the three pistons together so that they act as one (Figure 4-19A).

To make the bottom and center horizontal segments, place repeaters directly facing into the back of the sticky pistons and place a redstone torch next to the middle repeater with redstone dust behind the other two repeaters, which should activate those six sticky pistons (Figure 4-19B).

For the vertical segments, place redstone dust on the middle block behind the sticky pistons, and a redstone torch under that block (Figure 4-19C). You can see how all of this looks in Figure 4-20.

Figure 4-18   *The face of the seven-segment display*

A

B

C

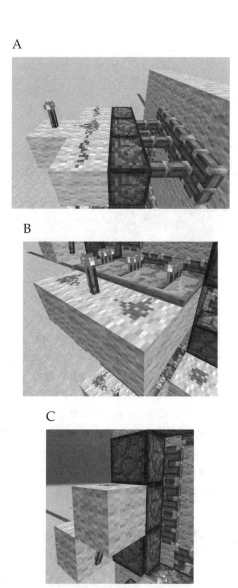

Figure 4-19 *Piston arrangements for horizontal and vertical segments*

Connecting the display to each of the output repeaters can be difficult, as I mentioned earlier. A certain amount of trial and error will probably be necessary before you manage to get all the signal lines between the piston tape and the display (Figure 4-21).

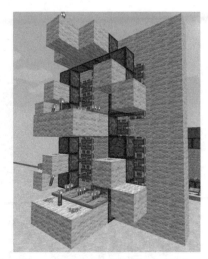

Figure 4-20 *The back of the seven-segment display with the pistons in place*

But once you've managed that, there's only one part left to make the display functional. Although connected, it won't actually do anything when you click the Cycle button for the piston tape. To fix this, we're going to replace certain blocks in the ring with a transparent block, for example, glass, so that the signal won't be able to get through, therefore turning certain segments off. To do this, find the segment you need to turn off, trace it back to the place in the piston tape, and replace the solid block there with glass. Once you've completed a layer, click the button and it should go to a blank layer on the tape so the display is blank, or an 8. You can do this until you

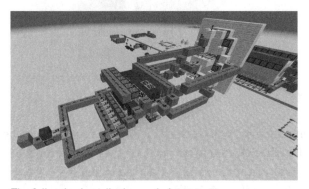

Figure 4-21 *The fully wired-up display and piston tape memory*

Figure 4-22 *Changing the data blocks of the piston tape*

have the numbers 0–9. You will find it easier to change the blocks on the top or back of the piston tape and then rotate it, than to try and get in among all the wiring. Figure 4-22 shows the top of the piston tape where the "data" blocks are easily accessible.

You've now made a functional seven-segment display! You can extend this to have two or more digits by making an identical piston tape, complete with order of numbers in the layers, and display. On the first one, extend the piston tape so that you have eight rings instead of seven. On this final ring, change all of the blocks to glass except for the block on the same layer as the zero for the first display. Instead of pointing the repeater inside toward the display, point it the other way. Put this line into the input of the second piston tape so when the first display gets to zero again, the digit on the second one will increase. You can repeat this as many times as you need digits.

Summary

In this chapter, we've learned how to build some of the more complex redstone contraptions, such as piston doors, BUD switches, and displays.

However, there's now only one chapter left on redstone, which includes some of the lesser known, quirky aspects of redstone, the miscellaneous redstone. After that, we'll be moving on to server hosting and modding!

5

Miscellaneous Redstone

In this chapter, we'll cover some of the quirkier parts of redstone and those parts that do not get used so often. However, everything in this chapter is still good to know as you'll probably want to use it at some point.

Some of the things we'll look at are: minecarts and rails, water physics in the game, hoppers, and command blocks, and you'll even learn how to make a TNT cannon.

Rails and Minecarts

Minecarts and rails, for quite a while, were almost always used solely for transport. However, with some fairly recent additions to the game, they became far more useful in redstone terms.

Minecarts are a kind of vehicle used in Minecraft. However, to travel, they must have rails (known as *mine tracks*) on which to run.

Lovers of model railroads can have a lot of fun with the various types of rails and minecarts. Using the different types of rail, you can start and stop minecarts as well as triggering other actions as minecarts pass over the rails.

In this section we will concentrate on the use of minecarts and rails with redstone. You will find very good information on building more conventional railroads here: http://www.minecraft101.net/redstone/minecarts-and-railroads .html.

Rails

There are currently four different types of rails in Minecraft: normal, activator, detector, and powered rails. All rails can be placed on opaque, or nontransparent, blocks, and mobs cannot go on rails unless they are pursuing the player.

The crafting recipes for all four types of rail are shown in Figure 5-1. Figure 5-1A is normal rail, B is activator, C is detector, and D is powered.

You can also often find sections of unused rails in abandoned mine shafts that you can take and use elsewhere.

Normal Rails

Even normal rails can be fairly expensive to build if you have very little iron; however, you do get 16 rails per 6 iron ingots and stick. As with all rails, they will automatically connect to adjacent rails. Normal rails are the only type of rail that will go around a corner, so they are often used as joining rails between the other types. As you lay the track, it will automatically be routed around bends if you place a rail at right angles to another.

Activator Rails

Activator rails only have one purpose, to activate TNT and hopper carts. We'll go into more detail about TNT and hopper carts later in this chapter. When a TNT cart passes over an activator rail, the fuse on it is set. The faster the cart is going when it goes over the top of the rail, the bigger the explosion. When a hopper goes over the top of an activator rail, it will disable the hopper cart. The activator rail only functions like this when it is being powered; if it is not powered, it acts like a normal rail.

A   B   C   D

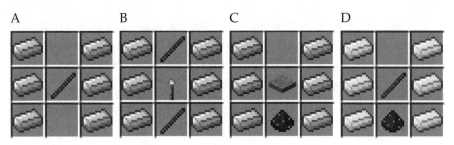

Figure 5-1 *The crafting recipes for the different types of rails*

To make sure that an activator rail is powered, you will need to use redstone. Usually a redstone torch is placed next to or above the rail. For more detail on how to power things with redstone, look at Chapter 2.

Detector Rails

Detector rails are similar to activator rails, but more generally useful because, rather than simply triggering TNT explosions, they activate rather like a redstone switch.

Normally, they act as standard rails, except that when a minecart goes over a detector rail, it outputs a redstone signal. It does not need a comparator, and no matter what type of cart it is or what is in the minecart, it will always output a signal strength of 15 as long as something is on it.

Powered Rails

Powered rails are either on or off. When powered rails are not powered "on," they will stop a minecart unless there is another "on" powered rail after it. When a powered rail is supplied with a redstone signal, it will light up and boost the cart's speed to a maximum of 8m/s (meters per second). When powered rails are placed adjacent to each other, they will act similar to redstone in that you only have to power one end, but the next eight rails will also be powered. There is no advantage to be had in terms of speed or anything else by powering adjacent rails until you have reached the eight-rail power range. It takes three powered rails to boost a cart up to maximum speed.

Minecarts

There are five different types of minecarts currently in Minecraft. The minecarts differ both in appearance and in how they behave when in contact with different types of rail.

The different types of minecart are

- Normal minecarts

- Chest minecarts

- Furnace minecarts

- Hopper minecarts

- TNT minecarts

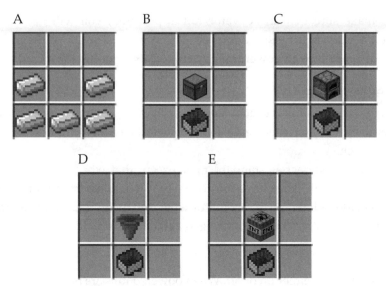

Figure 5-2 *The crafting recipes for the different types of minecarts*

All of them travel on rails, but only some of them react to certain rails. All of the crafting recipes are shown in the order above in Figure 5-2.

Normal Minecarts

To place any minecart, right-click with it on any rail. To get in the normal minecart, as a player, just right-click on it while it is placed, and press the LEFT ARROW-SHIFT key to exit. Generally, minecarts are faster than walking (4.3m/s) and sprinting (5.6m/s), and they travel at 8m/s (or 8 blocks per seconds) when on activated power rails. In addition to players riding minecarts, mobs, such as zombies, cows, or villagers, can also be "picked up" by minecarts and transported by them to other places if you want to put one on display or make a villager farm, and so on, which is often easier than using a water stream.

Minecart with Chest

The chest minecart functions exactly the same as a normal minecart except that when you right-click it, instead of riding it, you will open a chest inventory. This is the standard single-chest inventory so there is not a lot of space in it, but if you use several at once, it can make transporting items a lot easier.

In fact, you can build yourself an entire train to contain your possessions.

Minecart with Furnace

A furnace minecart is like the engine of the train. When right-clicked with coal or charcoal, it will push any minecarts in front of it, making them a good alternative to powered rails if you're short on gold. Sixty-four coals will power a furnace minecart for over three hours, which makes them ideal for pushing mobs on a makeshift rail line that doesn't have any powered rails.

Minecart with Hopper

Hopper minecarts are simply a minecart with a hopper, but they behave slightly differently than normal hoppers. A hopper minecart will pick up any items on the track it's running along, items above it, and items near the minecart. They will not automatically deposit items into containers below unless there is a hopper directly below it. Activator rails will also lock or unlock hoppers if they are powered. When a hopper is locked, no items can travel in or out of it, meaning it will no longer automatically pick up items thrown at it.

When trying to get an output from a hopper minecart with a comparator, you will need to place the hopper minecart on top of a detector rail with the comparator facing away from the hopper (Figure 5-3).

Minecart with TNT

TNT minecarts are simply used to blow stuff up, and there are a variety of ways to trigger them, such as: an activator rail, a derailment, or a drop of over three blocks, if it is destroyed mid-motion by the player, fire, lava, or an explosion, and finally if it turns a corner with either an entity or a solid block located adjacent to the track on the corner.

Figure 5-3 *Getting an output from a hopper minecart using a comparator*

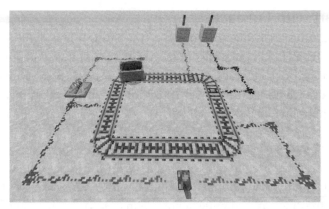

Figure 5-4 *A minecart and rail example*

Minecarts only derail when there are no more rails in front of them, so they fall onto the next block in front. Minecarts can still be pushed when derailed, but they still have to be placed on rails in the first place and they are harder to push when they are not on rails.

A Minecart Example

Figure 5-4 shows how you can use a combination of powered and normal rails with some redstone parts.

The lever will control the power to the powered tracks, allowing you to stop the minecart. To get the cart to start, you do need to give it a push.

As the cart travels around the track, it will cross two detector rails. As each is crossed, it powers the redstone going to the two torches at the back of the contraption. This approach can be used, with a suitably large track, to make a clock.

Liquid Physics

There are two different liquids in Minecraft: lava and water. Some of the behavior of lava and water is a bit unusual, but because of this, they can also be quite helpful. Both water and lava can be contained in a bucket by right-clicking the source block (the source block is still, whereas normal water/lava blocks look as if they are flowing).

Once you have lava or water in a bucket, you can place it wherever you like by right-clicking with it. It is placed like a normal block, so it needs to be placed while aiming at a block within reach. Water and lava buckets do not stack, so it can be very annoying if you want to build a lake and you don't have a nearby water source to just refill the buckets on your hot bar. In addition to lakes, you can build waterfalls or dams or even just use them for decoration.

Lava

Lava burns and damages players or mobs that come into contact with it. It will also burn items dropped into it, making it a useful garbage chute, or an annoying death-trap where you lose everything. If you use a fire resistance potion, then lava will not affect you. Lava can also be used with water to produce stone, cobblestone, and obsidian. To make stone, lava needs to be running vertically onto still or moving water. To produce cobblestone, lava has to run horizontally into a still or flowing water block. Finally, to make obsidian, water has to flow onto any side of a lava source block, consuming the lava source block in the process. You can make renewable cobblestone and stone generators but not renewable obsidian generators without taking advantage of bugs that don't always work. Lava also flows much more slowly than water, and it is not possible to construct an unlimited lava source as you can with water. If you want to find a lot of lava, then the nether is full of it and usually contains massive lakes of lava, as shown in Figure 5-5.

Figure 5-5   *An example of a huge lava lake in the nether*

Water

Water flows much faster than lava, and it is also fairly easy to make unlimited springs with water. The simplest way to do this is to either make a 2×2 hole or a 3×1 hole. Both only require two buckets of water. Place them in the corners of the 2×2 hole or the sides of the 3×1 hole, but you can only take the center block with the 3×1 design (see Figure 5-6). This works because in Minecraft, if a water block is adjacent to two or more source blocks, it will become a source block as long as there is either a solid block or another water block underneath it. When a player or a mob falls in water, they take no fall damage. This can be especially useful if you make a 1×1 hole, and place a sign in it and then a water above the sign, so the water will not flow past the sign. This creates a water block that a player, or mob, can fall through without taking any damage as long as it's not too far off the ground. Water is also

A

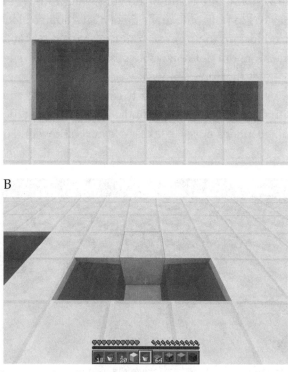

B

Figure 5-6   *Two examples of unlimited water sources*

very useful because it will absorb all explosive block damage. If an explosion occurs in water, it will not break any of the surrounding blocks; however, it will still launch and damage players and/or mobs.

Hoppers

Hoppers are a fairly recent addition to the game, but also a very useful one. When there is an item on the ground near a hopper, it will pick the item up and place it in its inventory until it is taken elsewhere. A hopper can also be used to pull items out of inventories and place them in other ones. When they are placed, the bottom part will face the side of the block you placed it on (except upward). Hoppers can also be locked by powering them with redstone, but that's not the only way they interact with redstone. You can use a comparator to get an output with them, which depends on the number of items. The exact math for this is explained in Chapter 2. They can be used to transport items; however, they are quite expensive to build in Survival mode, so you don't use them very often for that purpose.

In addition to being used to transport items, they can also be used for redstone purposes, in particular, a type of redstone clock. If you place one hopper facing any direction, then place a hopper facing into the first hopper, and then remove the first hopper and replace it so that it is pointing into the second hopper, the hoppers will be pointing into each other. Now, just place a comparator out of both of the hoppers pointing into a block. Then place redstone on the other side of the blocks with sticky pistons next to them. Finally, just place a redstone block in between the pistons (shown in Figure 5-7). To turn it on/off, you can just lock one of the hoppers, by

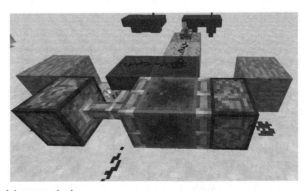

Figure 5-7   *A hopper clock*

powering the green line. Also, the length of time of the clock depends on how many items you put in; the more items, the longer the time it stays on/off.

One hopper feeds the other until it is empty; then they swap over, and this causes a tick. You can watch this process if you right-click on the hopper and watch the numbers in the inventory decrease. Note that you will need to right-click on the side of the hopper because the redstone dust will stop it opening from above.

Command Blocks

Command blocks are very useful to map makers; however, they are not obtainable without cheats because you could use them to cheat, so they are only available in Cheat mode. To get one, first you will need a world with cheats enabled (set when creating the world on more options), and then you need to press the T key on your keyboard. This will bring up multiplayer chat. In the chat, type **/give** *playername* **137 64**. Instead of *playername*, type in your Minecraft username. Alternatively, you can type **/give** and then press the TAB key and it will put in your name for you. The number 137 is the Item ID for command blocks, and 64 is how many it gives you, which can be changed to however many you need.

Command blocks can execute any server commands, which we'll look at more later, as well as some other commands. It will not execute the command unless the command block is being powered. They can also detect where the nearest player to the block or a specific location is, as well as being able to select all players, or random players. They can also deal potion effects, spawn blocks, play sounds, output redstone signals based on true or false parameters, and change the time of day. A full list of all of the commands and arguments can be found at http://tinyurl.com/CommandBlocks.

Player Traps

Player traps are usually completely hidden redstone contraptions that will kill the first player that triggers them. Lava is usually used because it is more compact, but it will destroy all of the items that person is holding, so making a big pit so that they die of fall damage will allow them to retrieve their items. Here, we'll go over two different player traps; one uses command blocks, so it would normally only be used in mapmaking, and also a second redstone version that is not completely hidden but is pretty close.

People use mapmaking to design worlds for the enjoyment of others, often incorporating traps and challenges for their guests.

Command Block Player Trap

Using a command block to kill a player is very easy. Although the **/kill** command cannot be used by command blocks, there are other ways to annihilate an opponent. In this example, we will make the command block search for any player in a four-block radius (the area is a sphere with a radius of four using the command block as the center) and teleport them to a hidden pit of lava. Again, alternatively, you could have it teleport the player really high into the air so that they die of fall damage and don't lose their items. To do this, first place the command block and then type this in the console command box:

```
/tp @p[r=4]  -408 49 775
```

The **/tp** is the teleport command, **@p** means it will be the nearest player (this could be replaced with a specific username), **[r=4]** means it searches a sphere with radius of four around the command block (which could also be replaced with whatever value you want), and the last part is the coordinates that it teleports the player to, which you'll replace with the coordinates for wherever you want it to teleport them to.

If you don't know how to find coordinates, it's actually very simple in Minecraft. First, you will need to press F3. Depending on your keyboard (especially if it's a laptop keyboard), you might need to hold the FN, or function, key, and then press F3 while you are holding it. This should bring up text all over the screen (see Figure 5-8), and the bit we're looking at, indicated in Figure 5-8, is x, y, and z all followed by three numbers.

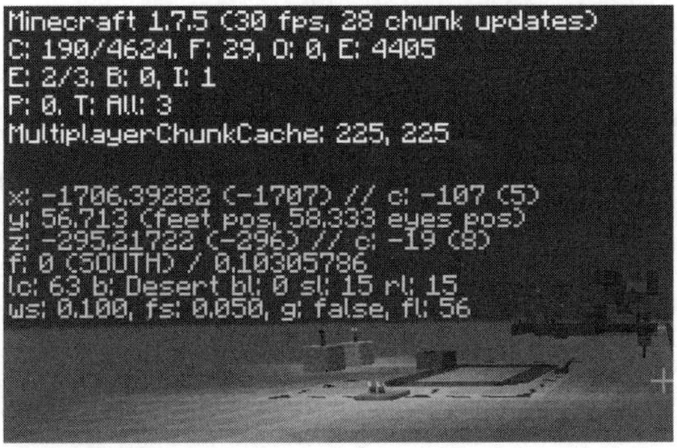

Figure 5-8 *Finding your coordinates*

The x and z are the horizontal axes (forward and backward, left and right) and the y coordinate is the vertical axis (up and down). Each whole one is one block, or one meter. The x and z coordinates can both be negative or positive numbers; however, the y coordinate cannot (if the player is teleported to a y coordinate of one or lower, then they will fall out of the world, which will also lose all of their items like lava). It is also worth noting that the Y level is your Minecraft character's head level, so two blocks above the floor that your character is standing on. To get the coordinates for where you want to teleport them, go and stand on top of where you want them to go, and bring up your coordinates. I recommend either writing them into a file on your computer or just writing them on a piece of paper until you need them again. Of course, in addition to using them for this player trap, coordinates also work in Survival Minecraft, so if you get lost, or find a really interesting place, you can write down the coordinates and go in different directions until you start heading toward your desired coordinates.

Finally, click the Done button and then set up a clock with its output powering the command block. Every time the clock pulses, it will run the command, so a faster clock means it will check more often, but a lot of fast redstone clocks can cause significant lag. See Chapter 3 for more information on using clocks.

A Redstone Player Trap

It is also possible to build player traps that use redstone and block updates, making them buildable in Survival Minecraft. It is very difficult to build a completely hidden player trap, but if it's a low-light area, you could just use tripwires to explode TNT or remove the floor, and so on. In this example, we are going to use the redstone ore block, which needs a pickaxe with the silk touch enchantment to get in Survival Minecraft. When a redstone ore block is walked over, or right-clicked on, it changes into a different block that has red particle effects and glows. We are going to make sticky pistons, which, when updated via the redstone ore block, will retract, taking the floor with them and making the player fall into a pit of lava.

First, you need to place a two-block-wide line of redstone ore blocks that form the floor of your corridor. To make it look less suspicious, just place some carpet on top of the blocks (shown in Figure 5-9).

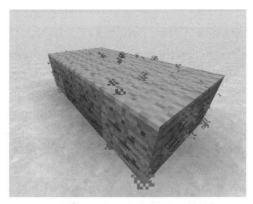

Figure 5-9 *A redstone player trap under construction*

Next, place sticky pistons behind one block behind all of your redstone ore blocks. Then, place any block one block up and behind (so it should be diagonal; see Figure 5-9). On all of those blocks behind the pistons, place levers and turn them all on. When you place one block next to any of the pistons on both sides, all of the other pistons should all update and extend. Hollow out a pit about three blocks deep underneath the redstone ore blocks and fill it with lava. All that's left now is to wait until the redstone ore blocks are back to the original state (not glowing or emitting particles), and then remove all of the levers. Only remove the levers after you have finished placing all of your other blocks if you're putting it into a wall or if you're building a wall around it.

The final trap is shown in Figure 5-10.

Figure 5-10 *The finished player trap*

TNT Cannons

TNT cannons don't really have any practical use, but every now and then it's fun to just blow stuff up. The 1.5 update to Minecraft changed dispensers' behavior quite a lot. Dispensers will dispense any item(s) placed in their 3×3 inventory when activated by redstone. The maximum any dispenser can hold is nine stacks of 64, or 576 blocks. One thing the update allowed is the ability to place entities from a dispenser, in this case, primed TNT, which previously would have just dispensed the item onto the ground, which is instead now done with droppers, which drop any item(s) in them when activated by redstone. A block of TNT inside a dispenser, when activated by redstone, will place the primed TNT block (which falls like gravel or sand) in front of it. This means you can have up to nine stacks of TNT before it has to be restocked, or even more if you put a hopper on top connected to a chest.

You can see an example of a variable TNT cannon in Figure 5-11. The variable is how far the TNT goes, the left-hand button is full strength, and the right-hand button is half strength. This works because the repeater only allows the redstone signal to go one way, so you're either firing one side or both sides. The repeaters going into the TNT that actually gets launched can be tweaked so that the less delay on the repeaters, the less time the TNT has to hit the ground before it explodes. If you want to make it fully automatic, then you can connect a fairly big delay (at least about 48 redstone ticks/12 repeaters at full delay) to either one of the buttons.

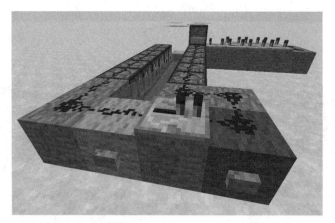

Figure 5-11   *A semiautomatic TNT cannon*

You can probably see the craters in the distance behind the contraption. Remember that all these builds are available to be explored (and fired) in the example world.

Summary

We've learned about some of the weirder, or sometimes less useful, parts of redstone, such as: minecarts, rails, some more components (command blocks and hoppers), how liquids work in the game, and even how to make a TNT cannon.

This is the last chapter on redstone, so now we can move on to some different technical aspects of the game. First, we'll have a quick look at how to set up a server so that you can play with other people from around the world in one shared Minecraft world. After that, the rest of the book will be all about mods, installing them, playing with them, and even making them!

6

Server Hosting and Tools

So far everything we've covered has been based on using Minecraft in single-player mode. But everything we've covered can also be made and used in multiplayer mode.

To use Minecraft in multiplayer mode, you will need to connect to a server. In this chapter, you'll learn how to set up your own server on a LAN (Local Area Network) if you and the person you want to play with are on the same network. You will also learn how to use Hamachi or port forwarding, so that you and your friends at other locations can play together in the same world over the Internet.

This chapter also looks at some tools that are available for Minecraft, in particular the Minecraft World Editor MCEdit. And for Raspberry Pi users of Minecraft, we will briefly explore the use of the Python interface to the Minecraft server.

Server Hosting

In this section, we will discuss various options for hosting servers. In addition to starting a new world to serve, you can also copy a "world" folder that you created in single-player mode to the server location to share it with others.

CAUTION   *I only recommend hosting servers if you have an unlimited data usage plan with your broadband provider; otherwise, it can get very expensive. Hosting Minecraft servers, especially if you have a lot of people on them, can also slow down the broadband for everyone else in your house connected to it, so it's a good idea to check with everyone else on your network to see if that's OK before hosting servers.*

Server Hosting on a LAN

The easiest way to play Minecraft with your friends is on a LAN. A LAN, which stands for Local Area Network, only works when everyone you want to play with is connected to the same home network (modem/router or home hub). You can be connected to the home hub by Wi-Fi or by direct Ethernet connection.

Playing on a LAN world is far easier than using Hamachi or port forwarding, but everyone has to be connected to the same network, meaning they'll have to be in the same house. With a LAN, there will be one host computer that everyone else is connected to. The host is usually the most powerful computer, so use whichever system runs Minecraft the best, because it will have to run Minecraft in addition to being a server. Once you have picked the host computer, go into the world you want to play on, which can be any world, press the ESC key, and then click Open to LAN. This should bring up a screen with a couple of options, such as the game mode everyone will be playing on and also if cheats are allowed or not. When you have finished this, click the Start LAN World button. The chat on the left of the screen, which can be accessed by pressing T on the keyboard, should say something like "Local game hosted on port 50582." The world is now ready to join.

To do this, anyone on the same network can go to the Minecraft main menu and then select Multiplayer. It should automatically find the LAN server for you to join, but if it doesn't, there is a relatively easy solution. To do this, you will have to add a new server, and put the IP address as the internal IP address for the host computer. If you are using Windows, go to the host computer, press and hold the WINDOWS key (usually found between the ALT and FN key on the keyboard) and then press R. This should bring up a run window. Type in **cmd** and then press ENTER. This should open up a command prompt. It may look scary and like the kind of program a hacker would use, but it's built into Windows and we'll simply be using it for diagnostic purposes. In the window, type **ipconfig**, or to do this on a Mac or a Linux system, open up Terminal and enter **ifconfig** instead of **ipconfig**. Depending on how your Windows environment is set up, you might need Administrative privileges to do this, but most of the time this is not needed. This should bring up a list of different network adapters and various odd-looking numbers. Scroll up or down until you find the subheading "Wireless LAN adapter Wireless Network connection" or "wlan." There might not be a 2 at the end, but as long as underneath it looks like the example in Figure 6-1, it should be fine. We're looking for the IPv4 address, which is given as 192.168.2.3 in our example and is circled. Almost all of these addresses start with either 192.168. or 10.0.0. You'll want to copy this number down somewhere or just put it straight into the IP address box on the Minecraft Open to LAN dialog. After this number, type a colon (:) and then the port number that we saw earlier in the chat. For example, mine would be **192.168.2.3:50582**. You should now be able to connect to the server.

Server Hosting Using Hamachi

If the people you want to play Minecraft with are not on the same LAN, then you can use a piece of software called Hamachi to allow your friends to connect to your server over the Internet. Hamachi is also available for Mac OS X as well as Windows, but for Linux users, you can skip this part and go straight to port forwarding. You do not have to use Hamachi to use a server over the Internet; see also the next section on port forwarding.

Hamachi is a program that makes running a Minecraft server easy, but it comes with limitations; for example, you can only have a maximum of five players connected to your server, you can't change the IP address that players have to type in to play on your server, and you have to have an extra program open, which can make your server lag more.

First of all, you will need to make a folder, wherever you want, called whatever you want ("myserver" might be a good name), which will contain all of the files and folders for your server. Then, go to www.minecraft.net/download and download minecraft_server.1.7.2.exe under the Multiplayer Server subheading, and place the application into the folder you just made.

A

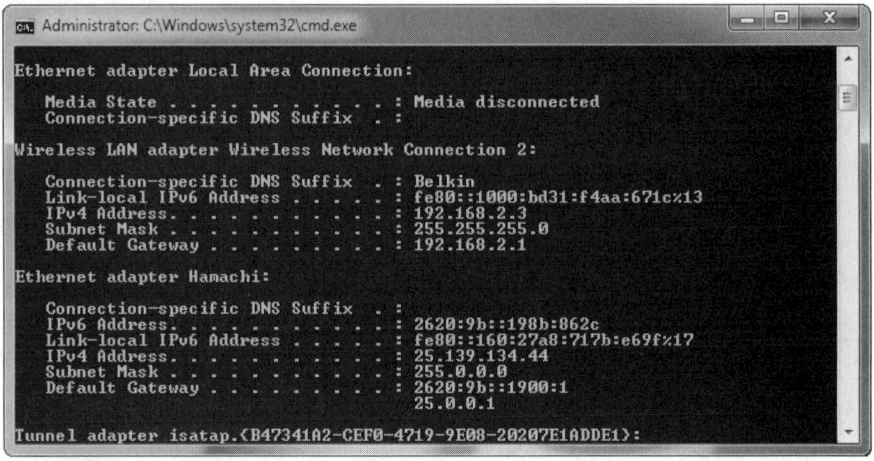

```
Administrator: C:\Windows\system32\cmd.exe

Ethernet adapter Local Area Connection:

   Media State . . . . . . . . . . : Media disconnected
   Connection-specific DNS Suffix  . :

Wireless LAN adapter Wireless Network Connection 2:

   Connection-specific DNS Suffix  . : Belkin
   Link-local IPv6 Address . . . . . : fe80::1000:bd31:f4aa:671c%13
   IPv4 Address. . . . . . . . . . . : 192.168.2.3
   Subnet Mask . . . . . . . . . . . : 255.255.255.0
   Default Gateway . . . . . . . . . : 192.168.2.1

Ethernet adapter Hamachi:

   Connection-specific DNS Suffix  . :
   IPv6 Address. . . . . . . . . . . : 2620:9b::198b:862c
   Link-local IPv6 Address . . . . . : fe80::160:27a8:717b:e69f%17
   IPv4 Address. . . . . . . . . . . : 25.139.134.44
   Subnet Mask . . . . . . . . . . . : 255.0.0.0
   Default Gateway . . . . . . . . . : 2620:9b::1900:1
                                       25.0.0.1

Tunnel adapter isatap.{B47341A2-CEF0-4719-9E08-20207E1ADDE1}:
```

B

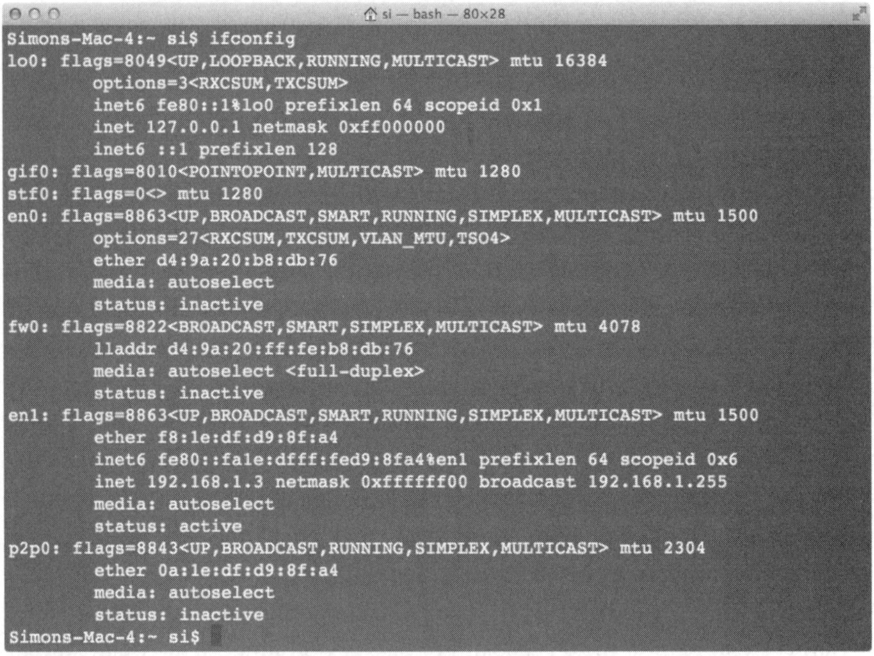

```
000                        si — bash — 80×28
Simons-Mac-4:~ si$ ifconfig
lo0: flags=8049<UP,LOOPBACK,RUNNING,MULTICAST> mtu 16384
        options=3<RXCSUM,TXCSUM>
        inet6 fe80::1%lo0 prefixlen 64 scopeid 0x1
        inet 127.0.0.1 netmask 0xff000000
        inet6 ::1 prefixlen 128
gif0: flags=8010<POINTOPOINT,MULTICAST> mtu 1280
stf0: flags=0<> mtu 1280
en0: flags=8863<UP,BROADCAST,SMART,RUNNING,SIMPLEX,MULTICAST> mtu 1500
        options=27<RXCSUM,TXCSUM,VLAN_MTU,TSO4>
        ether d4:9a:20:b8:db:76
        media: autoselect
        status: inactive
fw0: flags=8822<BROADCAST,SMART,SIMPLEX,MULTICAST> mtu 4078
        lladdr d4:9a:20:ff:fe:b8:db:76
        media: autoselect <full-duplex>
        status: inactive
en1: flags=8863<UP,BROADCAST,SMART,RUNNING,SIMPLEX,MULTICAST> mtu 1500
        ether f8:1e:df:d9:8f:a4
        inet6 fe80::fa1e:dfff:fed9:8fa4%en1 prefixlen 64 scopeid 0x6
        inet 192.168.1.3 netmask 0xffffff00 broadcast 192.168.1.255
        media: autoselect
        status: active
p2p0: flags=8843<UP,BROADCAST,RUNNING,SIMPLEX,MULTICAST> mtu 2304
        ether 0a:1e:df:d9:8f:a4
        media: autoselect
        status: inactive
Simons-Mac-4:~ si$
```

Figure 6-1   *An example of an internal IP address in a Windows command prompt, and a Mac terminal*

Next, you'll need to download Hamachi. Go to tinyurl.com/ LogMeInHamachiLink, check the Conditions of Use box under the Download Now button, and then click Download Now under the Unmanaged section. Once the file has downloaded, run it and follow the instructions to install Hamachi.

Now that you have Hamachi installed, you will need to launch it. Once you've launched Hamachi, click the Power On button as the application prompts you to do. Then you can choose a name for your account. After this, it will say "probing" and you will need to wait until it changes. It should now give you two options, Create A Network or Join A Network. Because we are going to be the server host, click Create A New Network. Make a note of the network name and password because you will need to give them to the people you want to join. You should now see your network on the main screen. At the top of the window, just above the account name you chose, you should see a number that looks something like 25.139.134.44, shown circled in Figure 6-2. Right-click on it, and then click to copy the IPv4 address. For now, we're done with Hamachi, and can go back to the server folder we made.

Figure 6-2 *Getting the IPv4 address from Hamachi*

First, just double-click the server application to run it. A window will pop up and you should notice that some new files have been created in that folder. For now, just type **stop** into the text box at the bottom right of the window and it should close down. The main file we'll be looking at is the server.properties file, which, when prompted by Windows, you should open with any text editor of your choice (you can use Notepad if you don't have any others installed). First, find the line in the file that reads "server-ip =" and then paste the IP address we copied from Hamachi. There are lots of other useful options here we'll go over later, but for now close the window.

Now, we're nearly done! All that's left is to run the server again and you can join it. To join your own server, click Multiplayer in the Minecraft main menu, click Add New Server, call it whatever you want, and then type **localhost** as the IP address. You should now be on your own server.

For other people to join, they will also need to have installed Hamachi, and click Join A Network instead of create one (or click on Network and then select the Join An Existing Network option from the drop-down menu). Here they'll enter the network name and password that you set earlier and then it should say that they have joined your network on Hamachi. You can either send them the IP address we copied earlier, or they can right-click on your network and select Copy IPv4 Address from there instead. Finally, all they need to do is add a new server in Minecraft, pick a name, and then paste the IP address in there, and they should be good to go!

Server Hosting Using Port Forwarding

If you are confident with network administration, you can use port forwarding rather than Hamachi.

Port forwarding is hardest to set up, but is better than using Hamachi because you can have one less program running, freeing up space for Minecraft to run more smoothly. Also you can have as many people on it as you want, but that depends on how fast your computer is. This method still uses the application file we downloaded from the Minecraft.net site in the Hamachi method, but it needs us to do something slightly different, which will be explained later in this section. For other people to join your server, we have to open a certain port on your network, which with Minecraft, is 25565. To do this, you need to connect to your home network and go to its configuration page.

To get to your modem/routers homepage, we need to find out your network's default gateway, which involves doing something very similar to finding your IPv4 address in the LAN section of the chapter. First, use the **ipconfig** (or **ifconfig** on a Mac or Linux) command that we used in the LAN section in the command prompt, and instead find the default gateway, which is two rows underneath the IPv4 address (circled in Figure 6-3). This will also, most of the time, start with 192.168. or 10.0.0. In this example, it is 192.168.2.3.

CAUTION   *Opening up ports on your home network can make it vulnerable to attack, so it's a good idea to revert any changes you've made when you are not using your network as a server.*

Now, go to your browser and in the URL bar, enter the default gateway. This will either take you to your router's homepage or a dialog will open asking for a username and password. This username and password will have been set or told to whomever set up the router, so you will need to ask them if you didn't set it up yourself. However, by default, the username is usually "Admin" and the password is usually either "admin" or "password." If it went straight to the homepage, it usually asks you for the password (sometimes a username as well) when you try to change something. Every make of router/modem has a different page, set out differently and with

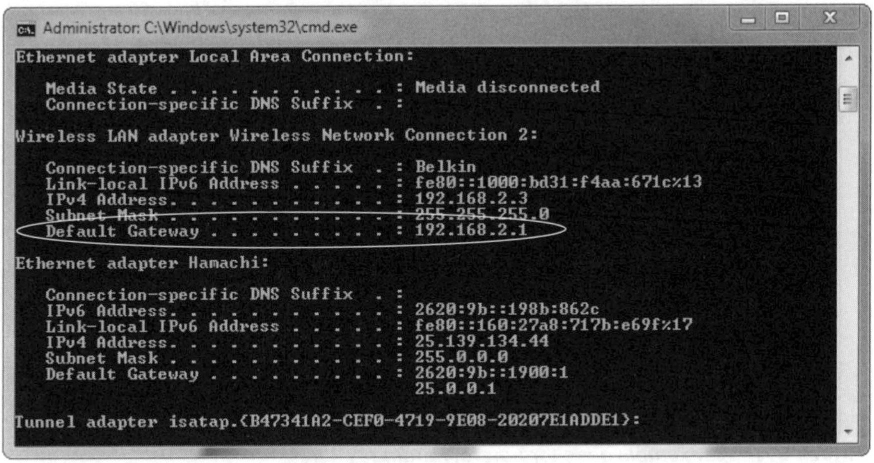

Figure 6-3 *Finding your default gateway*

everything called slightly different things, sometimes making it hard to find a specific feature such as port forwarding. The easiest way to find out where port forwarding is on your homepage is to search on the Internet for your router's make and model and where port forwarding is. Failing that, just go through the various categories until you find something that looks like Figure 6-4. If your router has port forwarding and port mapping, use port mapping, but be careful not to change any settings except for when we get to port forwarding, or you could mess up your router.

Once you have found this page, you will want to create a new port forwarding entry. If you are really lucky, there will already be an application type of Minecraft. If there is, then just add an entry of that type. If there is not, instead of selecting an application, there should be an option to type your

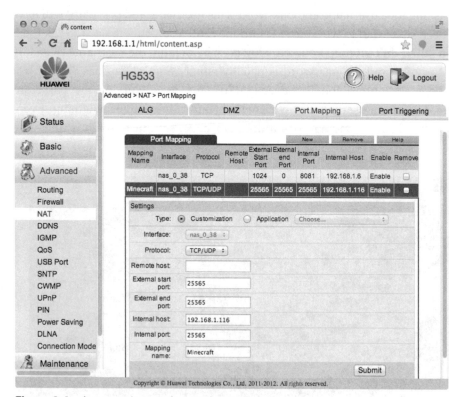

Figure 6-4 *An example page for port forwarding*

own, and just call it Minecraft. You need to set the protocol to TCP/UDP, or if you don't have this option, do it with TCP and then create a new identical one but with the protocol as UDP instead. If it asks for a remote host, leave it blank. Whenever you are asked for a port, internal/external start/end will always be 25565. Finally, set the internal host (or it could just ask for an IP address) to your internal IPv4 address, which we looked at how to find earlier. I recommend going into your router's LAN page, and finding an option for Lease Duration. Set this to "permanent," which will make sure your computer will always have the same internal IP Address when connected to that network. After this, just click Save, or Add, and you should see it get added to a table.

All we have to do now is create a new folder, and put the Minecraft server file from Minecraft.net/download in it. Run the file, and then type **stop** in the bottom left of the window after it has created a bunch of files in the folder. When the window has closed, open server.properties as a text file, go to the line that says "server-ip=" and put in your internal IPv4 address. Run the server one more time and you should be ready to start sharing!

To join your own server, add a new server in the Multiplayer menu, and put the IP address as your internal IPv4 address :25565. For example, if my internal IP address was 192.168.1.4, then I would put 192.168.1.4:25565. For other people to join, you will need your external IP address. The easiest way to find this is to just go to Google, and type **what's my IP**. It should just tell you at the top of the page, but if it doesn't, go to www.whatsmyip.org and it will tell you there. All your friends have to do is type in your external IP address as the IP for the server and they should be able to join as well.

Something to bear in mind is not to just give your external IP address to anyone you don't know, and to make sure that your home hub admin page does not allow remote administration. Most players have this feature turned off by default. You should also reboot your router once you've done server hosting. This is because people can use your external IP address to gain access to computers on your network, perhaps to install a botnet agent or look for personal information. This is pretty unlikely, but just as a precaution, only give out your external IP address while the server is running, and after you have closed down the server, reboot your router, which will give it a new external IP.

Server Properties Options

The server.properties file is used by both the Hamachi and port forwarding methods of doing multiplayer, so it is useful to know some of the things you can modify in this file to make a multiplayer setup work as well as possible.

This is the file with all of the different options for your server. Some of them are self-explanatory, whereas some of them are a bit more ambiguous. The first option, generator-settings, is for generating a specific super flat world. In single-player Minecraft, when creating a new world, if you select super flat as the world type, there is a Customize option. If you click Presets, you get a bunch of different options, which when clicked, indicate the type of super flat world Minecraft will generate. For example, the classic preset starts with a "2;" for the version; 7 is the block ID for bedrock, meaning there will be one layer of it; 2×3 means two layers of block ID 3, which is dirt; "2;" is the block ID for grass, and the semicolon means it is the end of the layers. Finally, 1 is the biome ID, which is a plains biome, and then "village" means that villages can spawn there. If you want more information on this, go to tinyurl .com/SuperFlatPreset.

The next option, op-level-permission, I would personally leave alone. However, you can change this to 1, 2, 3, or 4; one being the lowest level, and four the highest level of permissions. When someone is "opped" in Minecraft, they basically become an administrator, so they can do more than a normal player. The Op permission level indicates how much more they can do than a normal player. Op level 1 means that they can ignore the spawn protection rules (blocks around spawn cannot normally be broken, set by the spawn-protection option in the file). Level 2 ops can use the following server commands: **/clear, /difficulty, /effect, /gamemode, /gamerule, /give**, and **/tp**, and they can also edit command blocks. Level 3 ops can use a couple more commands: **/ban, /deop, /kick**, and **/op**, and. Finally, level 4 ops can use the **/stop** command to shut down the server.

The next interesting option is level-name=. By default, it is set to world. If you go into your server folder, you will notice there is a folder called world, which is the world folder that server uses. If you want to use a world you have from somewhere else, put the folder for that world into your server folder and change =world to =foldername, where *foldername* is the name of the folder with your world in. If you put the name of a world that isn't in your server folder, it creates a new world called that name.

Difficulty= and Gamemode= are other important options. Difficulty 0 is peaceful, 1 is easy, 2 is normal, and 3 is hard. Gamemode 0 is Survival, 1 is Creative, and 2 is Adventure.

The final option I'm going to go over in this chapter (a full list can be found at tinyurl.com/ServerOptions) is the motd= at the bottom of the file. This is the message of the day option, which when the server is added as a server, is the message it displays. This can be changed to anything you want; however, if it is over 59 characters, it might say "communication error."

Renting Servers

In addition to hosting your own server, you can also pay for dedicated servers that will run your server as long as you want and, most of the time, you can select how much RAM your server gets; the more RAM, the more expensive, but also the faster the server will run. Some examples of Minecraft server hosting sites are: mcpowerhosting.com, cubedhost.org, and www .aim2game.com. There are lots more all over the Internet, but these are just some of the ones that I know work well.

For very light usage (not suitable for redstone) where you only have a couple of people playing together, you can try setting up a server using Amazon's Elastic Cloud 2 service. The lowest tier of "tiny" is currently available for free (subject to certain limitations). You can find out more about hosting Minecraft on Amazon's Elastic Cloud 2 here: http://www.minecraftforum .net/topic/209252-amazon-ec2-server-setup-guide/.

MCEdit

MCEdit, as the name suggests, is a tool for editing Minecraft worlds. It is written in Python and available for use on Windows, Mac, and Linux computers. Download it from: http://www.mcedit.net/.

If you have ever looked at a contraption and thought with a sinking heart that you just need to make another ten of those, then you will love the idea of being able to copy and paste in Minecraft.

The MCEdit tool is an offline tool. That is, you do not run it while Minecraft is running. To be safe, just close Minecraft completely before starting MCEdit. Figure 6-5 shows the start screen of MCEdit.

Before you click the Load World option, it is a good idea to copy the world folder, in case you mess up the world. MCEdit looks for worlds to edit in the normal Minecraft single-player location, so if you wish to edit a server world, then you will need to copy it into your Minecraft world "saves" area.

Figure 6-6 shows this book's example worlds opened in MCEdit.

It can be quite tricky navigating in MCEdit as the program does not use the mouse to change the camera angle. Instead, use the I, J, K, and L keys to pan up, left, down, and right, respectively. In addition to the normal Minecraft navigation keys, the keys Q and Z will move you up and down. You may well have to move up if you find yourself underground. If you have an older computer, you will also notice that the program is quite slow to update.

The toolbar at the bottom of the screen gives you various options. Hover over these to get an idea of the things you can do with MCEdit. The leftmost icon is "Select." Try using this to select an area of the world. When you have done this, you will see options displayed as shown in Figure 6-7.

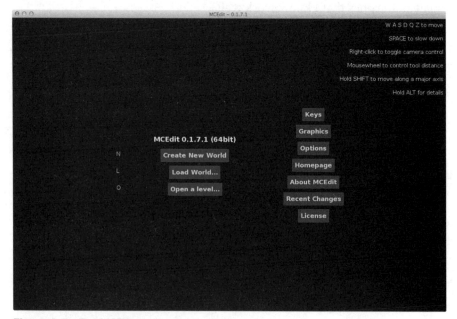

Figure 6-5 *The MCEdit main screen*

Figure 6-6   *The example world in MCEdit*

In this case, let's copy that contraption. Having done so, you can then paste the selection elsewhere in the world. If you just wish to move a contraption a little, then use the Nudge command.

When you have finished editing your world, do not forget to press CTRL-S to save the changes.

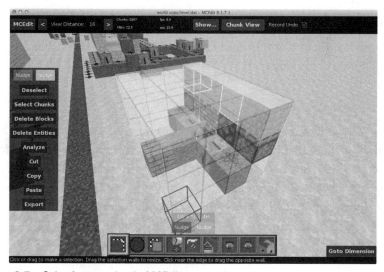

Figure 6-7   *Selecting a region in MCEdit*

Minecraft, Python, and Pi

The Raspberry Pi version of Minecraft has one very important feature that is missing from all other versions of Minecraft. That is, it includes a programming interface that you can use with the Python language. To make use of this interface, you really need a second computer on the same network as the Raspberry Pi that you can use to send Python commands or run Python programs on the Raspberry Pi without having to continually pause Minecraft.

Installing Minecraft on the Raspberry Pi

Minecraft is free on the Raspberry Pi. To install it, you need to enter a few commands in a terminal session. You will find instructions and the download here: http://pi.minecraft.net/.

The only tricky bit is actually downloading the file, as Midori and other Pi browsers can be unreliable at downloading files, so the most reliable way is to open LXTerminal on the Raspberry Pi and enter the following commands from your home directory:

```
$ wget https://s3.amazonaws.com/assets.minecraft.net/pi/
minecraft-pi-0.1.1.tar.gz
$ tar -zxvf minecraft-pi-0.1.1.tar.gz
```

This will create a folder called "mcpi" (Minecraft Pi).

Running Minecraft

You now need to make this your current directory and run Minecraft using the following commands:

```
$ cd mcpi
$ ./minecraft-pi
```

The Minecraft Launcher will start up and you will be able to create a new world. Note that the Pi edition of Minecraft is Creative mode only. Note that Minecraft will only be visible if you actually have a monitor physically connected to your Raspberry Pi. If you are connecting to your Pi remotely using VNC, then you will not be able to play the game.

Connect to the Pi from a Second Computer

Now that you have Minecraft running on the Raspberry Pi, you can use it in the normal way. However, if you want to use it with Python, you need to connect to the Raspberry Pi from a second computer. You only need a command-line interface, so follow the instructions at this Adafruit tutorial to get you connected from a PC, Mac, or Linux computer on the same network as the Raspberry Pi using "ssh": http://learn.adafruit.com/adafruits-raspberry-pi-lesson-6-using-ssh.

Using a terminal session connected to the Raspberry Pi, you can send commands and run Python programs that will change the "world" running on the Raspberry Pi, as you watch. Let's start by just sending a talk message to the screen of the Raspberry Pi.

Using an SSH terminal session, enter the following command to put you in the right folder for these Python experiments:

```
cd /home/pi/mcpi/api/python/mcpi
```

Then enter the command **python**:

```
$ python
Python 2.7.3 (default, Jan 13 2013, 11:20:46)
[GCC 4.6.3] on linux2
Type "help", "copyright", "credits" or "license" for
more information.
```

Now that Python is running its command line (see Figure 6-8), you will notice that the command prompt has changed to >>>. Now enter the following commands and you should see a talk message pop up on the Minecraft game on the Raspberry Pi.

```
>>> import minecraft
>>> mc = minecraft.Minecraft.create()
>>> mc.postToChat("Hello Minecraft World!")
```

The first two lines just set up the interface from Python to Minecraft. The final line uses "postToChat" to send the message. Try changing the message and sending a few more commands. Figure 6-9 shows the message appearing on the Minecraft screen.

Let's now try writing a short program that will build a staircase of blocks from the player's current position. Exit the Python prompt by typing **exit()** and then open a text editor using the following command:

```
$ nano staircase.py
```

```
pi@raspberrypi ~ $ cd /home/pi/mcpi/api/python/mcpi
pi@raspberrypi ~/mcpi/api/python/mcpi $ python
Python 2.7.3 (default, Jan 13 2013, 11:20:46)
[GCC 4.6.3] on linux2
Type "help", "copyright", "credits" or "license" for more information.
>>> import minecraft
>>> mc = minecraft.Minecraft.create()
>>> mc.postToChat("Hello Minecraft World!")
>>>
```

Figure 6-8 *Interacting with Python and Minecraft*

Type the following code into the editor window and then press CTRL-X and then Y and ENTER to save the file.

```
import minecraft
import block
mc = minecraft.Minecraft.create()

mc.postToChat("Let's Build a Staircase!")
x, y, z = mc.player.getPos()
for xy in range(1, 50):
    mc.setBlock(x + xy, y + xy, z, block.STONE)
```

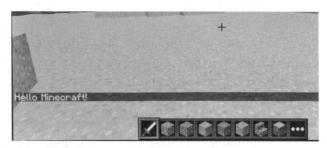

Figure 6-9 *A message received in Minecraft*

Figure 6-10   *A staircase created by Python*

To run the program, enter the following command:

```
$ python staircase.py
```

It will appear as shown in Figure 6-10.

Taking a look at the program, you can see that the first thing we do is use "mc.player.getPos" to find the current X, Y, and Z coordinates of the player. We then have a loop, which effectively counts up to 50. Each time it does so, "mc.setBlock" is used to create a block at the coordinates of the current player position, with both x and y coordinates having the current count value ("xy") added to them. Although this staircase is 50 blocks, you could increase that number to make the staircase extremely long.

Martin O'Hanlon has created a very useful tutorial and documentation for using Python with Minecraft here: http://www.stuffaboutcode.com/2013/04/minecraft-pi-edition-api-tutorial.html. It includes useful examples and documentation for all the Python commands you can use.

If you want to learn more about programming the Raspberry Pi using Python, then you may find the book *Programming the Raspberry Pi: Getting Started with Python* by Simon Monk useful.

Summary

This chapter has been all about servers and tools. We've learned how to host a LAN server with your friends if you're all connected to the same network, how to set up a server easily with Hamachi, how to set up a server using port forwarding so that you have more control over it, and we've had a quick look at some other Minecraft tools.

This chapter concludes this section of the book, which is all about vanilla, or un-modded, Minecraft. Now we're going to start taking a look at some of the existing mods for Minecraft such as ComputerCraft, where a computer is a single block, complete with a programming language so that you can make and execute your own programs, and also how to make your own mods that you can distribute.

7

qCraft

qCraft is a mod for Minecraft that tries to incorporate the basic principles of quantum physics in the Minecraft world. It was produced by Daniel Ratcliffe, Google, Caltech, and others. For more information, the official website for the mod is http://www.qcraft.org.

In this chapter, we will cover the basics in qCraft, from teleportation, to wireless redstone. It can take a while to get used to crafting some of the blocks for this mod, and you often need to know which direction you are facing in the Minecraft world, but once you get the hang of it, the possibilities are endless!

Getting qCraft

The quickest way to start playing qCraft is to install a mod pack and launcher such as the Tekkit mod pack used with the Technic Launcher. This Launcher replaces the normal Minecraft Launcher (you must still have the official Minecraft and a Minecraft account).

Download the platform from http://www.technicpack.net/download. It is available for Windows, Mac, and Linux. When you run the launcher, select the "Tekkit" modpack from the list of modpacks on the left-hand side of the screen (Figure 7-1). This modpack contains both qCraft mod and the ComputerCraft mod that we will be exploring in the next chapter. It also includes a whole load of other mods that are also worth exploring.

Figure 7-1 *The Technic Launcher*

Installing the Example World

All the example contraptions described in this chapter are set up in an example world that you can explore as you work through the chapter. In fact, it will make it much easier to follow the instructions if you download and install this.

The first step is to download the zip file of the world from:

https://github.com/simonmonk/minecraftmastery/tree/master/qcraft_example_world

To download the zip file, you need to first click the link "qCraft Example World.zip" and then on the page that opens, click the "view raw" link. Save the zip file on your desktop and unzip the file. You then need to copy the entire "qCraft World" folder into your Technic Launcher's saves folder.

The location of your saves folder will depend on which platform you are running Minecraft on. In both Mac and Windows, these files are hidden by default.

The easiest way to find the folder, which works on all platforms, including Windows, is to select Options from the main menu of Technic Launcher and then click Texture Packs and then Open Texture Packs Folder. Then navigate up one level and you will see the .tekkitmain folder, which also contains the saves folder.

When you have copied the qCraft Example World folder into the saves folder, you will find that when you start up Tekkit/Minecraft again, the new world will be there. You can open it and explore the world as you please. If you accidentally damage one of the contraptions that you are looking at, you can always copy the files again.

Quantum Dust

Quantum dust is used for the crafting recipes of all of the other items in qCraft. It is obtained through mining quantum ore, which is very similar to redstone ore in a lot of ways. It only spawns in the bottom 16 layers, it drops the same amount of quantum dust per block as redstone (4–5), and it can only be mined with an iron pickaxe or better. However, it only appears half the time per chunk making it a rare commodity.

One item that quantum dust is used for is quantum goggles. These are made by putting a glass pane on the left and right of a quantum dust. To use them, simply put the goggles in your helmet slot in your inventory and everything will turn slightly green. The goggles are used to find all of your quantum modified blocks (Observer Dependent Blocks, Entangled Blocks, and Quantum Blocks) in your world.

Essence of Observation

Essence of Observation is made by making a diamond out of quantum dust in a crafting table (see Figure 7-2A-C for the crafting recipes of all of the essences). Once you've got an Essence of Observation (EoO), you'll want to make an Observer Dependent Block (ODB). At first, the crafting recipe for this may seem a bit odd, but after you've made a couple of them, it's much easier. To start off, place your Essence of Observation in the middle of a crafting table. An ODB works like this: depending on where you look at the block, it will become a different block. For example, if you looked at it from

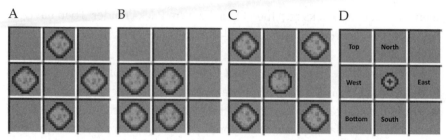

Figure 7-2   *The crafting recipe for Essence of Observation (A), superposition (B), entanglement (C), and the crafting recipe for an Observer Dependent Block (D)*

below, it could appear as dirt, but from above, it could be stone. Where you put your selected Minecraft blocks in the crafting grid dictates how it will change, as labeled in Figure 7-2D. If you leave any of the required spaces blank, then the block will turn into an air block when looked at from that side, therefore becoming invisible.

One way we can take advantage of this invisible-from-certain-sides feature is to make a very simple hiding place for crafting tables or chests, and so on. In fact, it only requires one ODB and some building blocks. First, you'll want to place your hidden block (in this example, we'll use a crafting table) and then create a wall one block wide in front of your hidden block, as in Figure 7-3.

Now, we need to make the OBD. How you make your OBD will depend on which direction your wall is facing. Press F3 and look toward the bottom left, and you should see which direction you're facing (interestingly, a Minecraft compass is completely useless at finding which direction you're facing, as it points to your spawn point). Next, look at your hidden block standing in front of your wall and note which direction you're looking at. In the crafting table, place an Essence of Observation in the middle, and place the same block that you are using for your wall at the opposite side to the direction which you previously noted (North is the box above the center, East is the box to the right of the center, South is underneath the center, and West is to the left of the center). For example, if my wall was facing the West, I would place my building block on the right of the center (East). Once you've done that, place the OBD in front of the hidden block and it should blend in with the wall. Now, to access the hidden block, simply move against the wall,

A

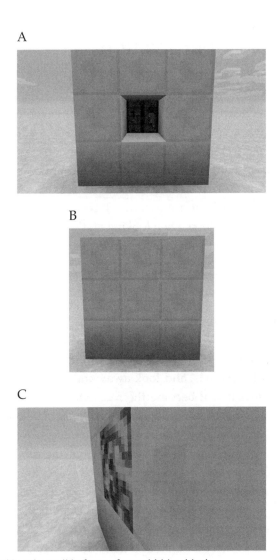

B

C

Figure 7-3   *Making the wall in front of your hidden block*

look away from it, and look back. The block should now have disappeared, allowing you to interact with the hidden block. To make it blend back in with the wall, face the block head-on, look away and look back again, and that's it!

Essence of Superposition

Essence of Superposition (EoS) is similar to Essence of Observation, but slightly different. It is crafted by making a 2×2 grid of quantum dust in a crafting space. The Essence of Superposition (EoS) when combined with a block works like an OBD except it doesn't always change depending on where you look; it is random. To craft a quantum block (a block using the EoS), you do the same as you did with the ODB except you use an EoS instead of an EoO (Essence of Observation). Instead of each direction being specific, it is each axis (left right, front back, and top bottom). So if, for example, you want a block to be either dirt or stone when looking at it from the East-West axis, put a dirt block on one side and stone on the other. As with ODBs, any directions not specified with a block will turn to air when looked at from the unspecified angle.

A very simple redstone randomizer can be made by making a quantum block. To do this, put a redstone block on either the West or East position, stone on the other West-East position, and another block type, for example, a quartz block, in all of the other positions. If you now place this block down, face it on the East-West axis, and look away and then look back, there is a random chance that it will become the redstone block, which will output a redstone signal. If it is truly random, the likelihood of you looking at it five times in a row without changing is one in 32. However, you quite often seem to get long streaks like this, which suggests that the mod seems to favor streaks more, but that could just be pure luck! However, if you do want your randomizer to take longer to get an output, you can chain a couple of these blocks together. If you make four of these Quantum Blocks, with a one-block gap between each one and a piece of redstone in front of all of them, then place a block with a torch on the side so that it looks like the five-input AND Gate from Chapter 3 except with four inputs instead of five. Figure 7-4 also shows how this could work, and you now have a randomizer that requires four blocks all to be on in order to get an output.

Figure 7-4   *A four-block redstone randomizer*

Essence of Entanglement

Essence of Entanglement (EoE) is made by putting one Essence of Superposition in the middle of the crafting table, and then putting quantum dust in each corner. When an EoE is combined with a block, it will give the player double the number of Entangled Blocks per each ingredient used. The crafting for an Entangled Block is either a Quantum Block or an Observer Dependent Block on both sides of an Essence of Entanglement. The resulting block will act exactly the same as the block used to make it except that when one changes, so does the other. So you place down one of your blocks, and then walk somewhere else, place a second one, and then look at one of them and then back at it so it changes, and then the other block will also change. You have groups of Entangled Blocks, so if you make identical blocks, they will not stack because they are in a different group, but they will still act the same.

We can actually use this property to make wireless redstone in a sense. If you create a pair of Entangled Blocks just like the ones shown in Figure 7-5A, place one down, and then place the other as far away as you want (just for this example, I'm going to place it about 10 blocks away). Now, if you look at one of the blocks from the North/South, then it will change to a redstone block, which outputs redstone on every side, so you can have a line of redstone next to any of the sides going to, for example, a door. When you face East/West and look away and look back, the block will change back to stone, as will the other one, closing the door again (see Figure 7-5B). You could instead make it so that it is invisible from every side except one, which would make it a lot harder to spot!

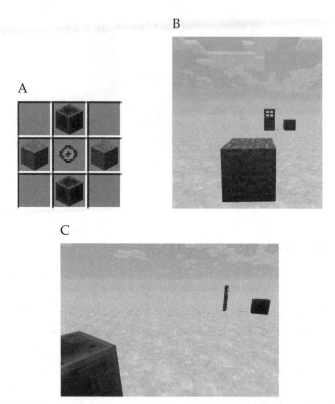

Figure 7-5 *The crafting recipe for the Observer Dependent Block to make the Entangled Block and the basic setup for wireless redstone*

Automated Observers

Being able to change a block just by looking at it is pretty cool, but if you want to make some more complex designs, you might need the Automated Observer. When a redstone signal is supplied, it will act as an observation update, meaning you can change a block without looking at it.

It is fairly simple to craft, one quantum dust in the center, surrounded by stone except for the bottom-middle grid, which is a redstone block. Now that you have your Automated Observer, you will need to know how to place it. Similar to a redstone repeater, it has a specific input and output side. When you place it down, the side facing you is the redstone input and the opposite side is the observing side. The block that you want it to observe must be directly next to the output (observing) side, and that's it. When a redstone signal is sent to the Automated Observer, it will change the block unless you are looking at it from a different angle that changes it, because as soon as it updates, looking at it from the other side will change it back again.

One use of this is to make a kind of redstone clock (see redstone logic gates in Chapter 3 for more information on redstone clocks) that can be very fast, half a redstone tick (a 20th of a second) up to however many repeaters you want to place, but the output of which can be accessed anywhere in your world.

First, you will need to make two lots of an Observer Dependent Block that is a redstone block on one axis (East-West, or right-left on the crafting grid for this example) and stone, or any other block for that matter, on the other axis (North-South or top-bottom) and the remaining slots can be any block you want also. Once you have two lots of this block (or more), you will want to make an entangled pair by placing them on either side of an Essence of Entanglement in a crafting table. You will also need two Automated Observers and some redstone dust. To start off, place down one of your Entangled Blocks with an Automated Observer pointing into one side, and the other Automated Observer into one of the adjacent sides (see Figure 7-6). It is important that you face the block while you place these to make sure that they are oriented correctly.

Next, you need to run a line of redstone dust connecting both Automated Observers, and also a line to the Entangled Block. All you need to do now is go to one of the sides where it will change to a redstone block, look away and look back, and hopefully you should just see the block changing animation and redstone pulsing very quickly (Figure 7-7 is the final build)!

To access the clock, simply place down one of the other Entangled Blocks from the same group and it too should be very quickly changed between blocks. The output is on any side of the Entangled Block. If you want to

Figure 7-6 *Setting up the Automated Observers*

Figure 7-7 *The clock is now ready to be accessed from anywhere.*

change the length of the clock, all you need to do is place repeaters between the redstone line from the Entangled Block to the Automated Observers and change the delays on them to whatever you want. You can also make different groups of the Entangled Block that are identical to the previous ones and set up the same with the Automated Observers, but, each time, you can set a different group to have a different delay, so then if you need a specific clock, you can just place down the corresponding block, saving you from having to make a new redstone clock for everything. However, if you're going to do this, you will probably want to name the different groups so that you don't have to figure out what each one does every time.

Quantum Computers

Quantum computers made by themselves do not actually do anything unless you use them in a structure, or in this case, if you entangle two of them, then they become useful (see Figure 7-8 for their crafting recipe). When you have a pair of entangled computers, we can start to do some really cool things. It is quite hard to set up, but once you have followed the instructions in this section, you should be able to teleport entire structures to a completely different location.

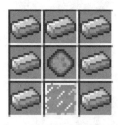

Figure 7-8 *The crafting recipe for a quantum computer*

First, after you've got your entangled quantum computers, you will need to make four slightly different Observer Dependent Blocks; however, you will need to have two of each type. The blocks need to work as follows: one block needs to appear as obsidian from every direction except North, where it appears as a block of gold, and another needs to do the same except every direction apart from South, one as East and one as West. It is worth organizing them in your inventory so that they are in an order so that you can tell which blocks are North, South, East, and West. You will also need at least eight blocks of obsidian, eight blocks of glass, and two blocks of ice.

The area of blocks to be transported is marked out with four beacons, but first, place down one of the entangled computers at the center of the area you want to teleport, and place an ice block on any side of it (including above or below it). Now, you will need to lay out your beacons, which have to be on the same level as your quantum computer. The furthest out they can be is eight blocks from the computer, so you will need to bear that in mind. I also recommend using placeholder blocks, such as dirt, to mark out where your beacons will go. The four different Observer Dependent Blocks we made earlier have to be laid out in a specific order. Press F3 (Function F3 on Mac) to bring the extra information as we did earlier, and look toward the bottom left to find out which direction you are facing. While standing between the quantum computer and one of your four placeholder blocks, note which direction you are facing. Now find the opposite direction Observer Dependent Block to that in your inventory, and replace the temporary block. For example, if I were facing East, I would place the block that appears as gold from the West there. Do this until you have four blocks that all appear as gold when you stand inside the area. Now, you will need to place obsidian on top of each beacon (you can have up to seven blocks above or below the Observer Dependent Blocks, meaning that 14 is the maximum amount if you do seven below and above) and then a glass block at the end of each tower of obsidian, meaning the maximum beacon height is 16 if you use blocks above and below the Observer Dependent Block including the glass (see Figure 7-9). All that's left now is to make an identical copy of that in the other location that you want to teleport your blocks to.

If you have done all this successfully, when you right-click one of your quantum computers, there should be a button labeled Energize. When you click it, the computer will transport all blocks within your beacons to the

Figure 7-9 *A quantum computer setup*

location of the other entangled quantum computer (see Figure 7-10 for an example). You can also use a redstone signal to energize the quantum computer if you wish. You can now teleport blocks with the power of quantum physics!

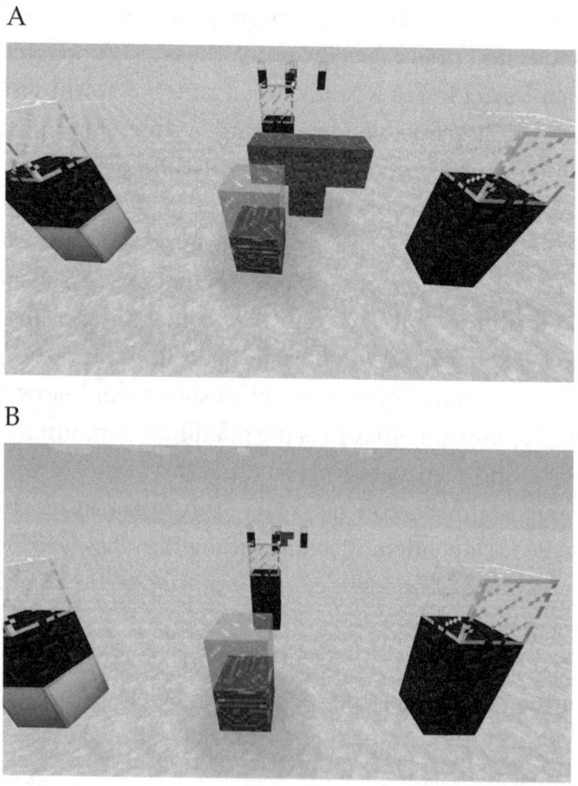

A

B

Figure 7-10 *Teleporting blocks with entangled quantum computers*

Quantum Portals

Being able to teleport blocks is pretty cool, but in qCraft, you can also teleport players elsewhere in the world, and even onto other servers. To do this, you will have to make a structure set out the same as a nether portal (http://minecraft.gamepedia.com/Nether_Portal) but with different blocks. You will need ten glass blocks, eight blocks of gold, sixteen blocks of obsidian, four Essences of Observation, an unentangled quantum computer, and a block of ice per portal; however, you need at least two portals to actually teleport somewhere else.

First, we need to make four Observer Dependent Blocks with gold and obsidian. Once again, you will need to bring up which direction you're facing with F3 again, and face the direction you want the portal to face (looking at the gap in the middle of the structure, not from the side). If your portal will be facing North/South, then when crafting the Observer Dependent Blocks, put four gold blocks on the North and South positions (above and below the Essence of Observation). Likewise, if your portal will be facing East/West, put the gold blocks to the left and right of the Essence of Observation instead. Simply place four obsidian blocks in each of the remaining slots, giving you four identical Observer Dependent Blocks.

Now we can start to build the actual structure. Place down one of the Observer Dependent Blocks on the floor, place two blocks of glass to the right or left of it, place another Observer Dependent Block, go up three blocks of glass, place the third Observer Dependent Block, go across two blocks of glass, place the last Observer Dependent Block, and then place three blocks of glass downward (see Figure 7-11). Finally, place your quantum computer adjacent to the portal and then the ice adjacent to the computer.

If you right-click on the computer, it should bring up an interface with "This Portal" at the top, "Destination Portal" underneath, an "On This Server" button underneath that (we'll get to that in a minute), and lastly an "Energize" (or "De-energize") button (Figure 7-12).

To edit the portal name and destination portal name, just click in the boxes and type. The "+" button next to the On This Server button denotes whether or not the portal you are attempting to connect to is on another server. If you do have a portal set up on another server, click the "+" button next to it and type in the server IP. Of course, before you can teleport, you will need another

Figure 7-11 *The layout of the quantum portal*

portal set up, so simply type its name in the destination portal field and then click Energize. The portal should then light up green, and if you step through, you should be teleported to the other portal. If you are traveling to another server, it will ask you if you want to keep your inventory; however, this can cause issues if you are not a server administrator on that server. Like the quantum computer block teleportation, you can use redstone to energize and de-energize it.

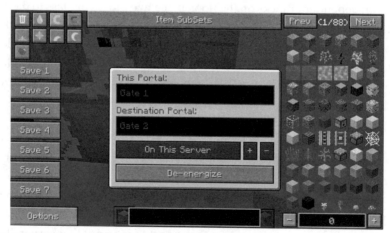

Figure 7-12 *The quantum computer's control panel*

Summary

We covered the very basics of this mod, but with other mods, and even on its own, there are all sorts of amazing contraptions to be built that we haven't touched upon on this chapter. The more you use this mod, the more quickly you realize just how powerful it can be. You could even try going through some of our earlier redstone contraptions and try to make them better or completely different with qCraft.

After you've mastered quantum physics in Minecraft, where do you go next? Well, fully functioning programmable computers in a single block sound pretty good. ComputerCraft implements computers, monitors, and even turtles into Minecraft, allowing for some really awesome things to be built!

Summary

8

ComputerCraft

In this chapter, we look at a fairly substantial mod called ComputerCraft.

ComputerCraft is a mod that allows you to make projects using computer blocks. You can program the computer in a language called Lua. The computer can also have inputs and outputs that connect to redstone devices. This allows the computer to control mechanical things in your Minecraft world.

In addition to computer blocks, there are also monitors and robots (called turtles) that can be programmed like a computer. You can also communicate between computers, making networks, as well as connect to the Internet and interact with web services such as Twitter.

An easy way to install ComputerCraft is to install one of the mod packs and launchers that include it, such as Technic Launcher and Tekkit (http://www .technicpack.net/download). You can find instructions for installing these at the start of Chapter 7.

Installing the Example World

All the example contraptions described in this chapter are set up in an example world that you can explore as you work through the chapter. In fact, it will make it much easier to follow the instructions if you download and install this.

The first step is to download the zip file of the world from:

https://github.com/simonmonk/minecraftmastery/tree/master/ computercraft_example_world

To download the zip file, you need to first click the link "Computer World .zip" and then on the page that opens, click the "view raw" link. Save the zip file on your desktop and unzip the file. You then need to copy the entire Computer World folder into your Technic Launcher saves folder.

The location of your saves folder will depend on which platform you are running Minecraft on. In both Mac and Windows, these files are hidden by default.

The easiest way to find the folder, which works on all platforms, including Windows, is to select Options from the main menu of Technic Launcher, click Texture Packs, and then Open Texture Packs Folder. Then navigate up one level and you will see the .tekkitmain folder, which also contains the saves folder.

When you have copied the Computer World folder into the saves folder, you will find that when you start up Tekkit/Minecraft again, the new world will be there. You can open it and explore the world as you please. If you accidentally damage one of the contraptions that you are looking at, you can always copy the files again.

A Computer Block

If you are playing in Survival mode, then you will need to craft a computer or an advanced computer. The recipes for these and all the ComputerCraft blocks can be found here: http://www.computercraft.info/wiki/index.php ?title=Recipes.

When you add a computer block to your world, it will look like Figure 8-1.

When you right-click on a computer block, you will see a console window like the one shown in Figure 8-2.

Figure 8-1 *A ComputerCraft computer*

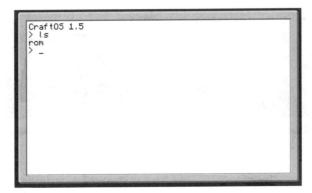

Figure 8-2   *A ComputerCraft console*

You can type commands at this window, as you can with the command prompt in Windows or the Terminal on a Mac or Linux computer. You can see that here the command **ls** has been typed. This displays the contents of the computer's hard disk (well, its imaginary hard disk, at any rate).

From this console, we can type programs into an editor, save them, and then run them.

Getting Started

Lua is a modern scripting language. It has a simple syntax and is an interpreted language, so you do not have to compile scripts before you can run them on your ComputerCraft computer within Minecraft. In addition to being able to write Lua programs in a file editor and then save and run them, you can also open a Lua console and type Lua commands directly into the computer's console.

We can start with a simple command that uses the computer to turn on a redstone lamp. Start by creating the arrangement shown in Figure 8-3.

Now right-click on the computer to open the console and type:

```
>lua
```

You can connect redstone to any side of the computer except the front. So in this case, we have a redstone lamp connected to the left side of the computer. By typing in the command shown in Figure 8-4, we can turn the lamp on.

Figure 8-3   *Attaching a redstone lamp to a computer*

Depending on the time of day in Minecraft, you may be able to tell that the lamp has come on. In any case, you can check by pressing ESC to leave the console. You should see the redstone lamp illuminated as it is in Figure 8-5.

Return to the console and press the UP arrow key to bring back the last command that you typed in Lua and change "true" to "false." When you press RETURN, the lamp should now turn off again. When you have had enough of the Lua console, you can return to the computer terminal by entering the command:

```
lua> exit()
```

The command **rs.setOutput** expects two values to be given to it inside the "(" and ")" separated by a comma. The first value is the side to use. This must be one of: "top", "bottom", "left", "right", or "back" and the second is "true" or "false" for on or off.

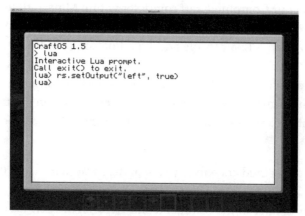

Figure 8-4   *Controlling redstone from Lua*

Figure 8-5   *The redstone lamp lit*

Now let's expand this example so that we have a redstone input as well as an output. The input side of the computer will be attached to a lever. When we activate the input by flipping the lever, a Lua program will cause the redstone lamp to flash. Figure 8-6 shows the same arrangement as Figure 8-3, but with a lever attached to the right side of the computer.

The logic for this arrangement needs a few lines of code to implement it. So rather than enter the code on the Lua command line, we are going to write a short Lua program in the ComputerCraft editor window.

Right-click on the computer to gain access to the computer's terminal and then type the following command:

```
>edit flash.lua
```

In the editor window, type in the program code, shown in Figure 8-7.

Hold down the CTRL key and then press s to save the file. You will see a confirmation message that the file has been saved to flash.lua. Hold down CTRL again and then press e to exit the program.

You can now run the program by typing the command:

```
> flash.lua
```

Figure 8-6   *A computer with redstone lamp and lever*

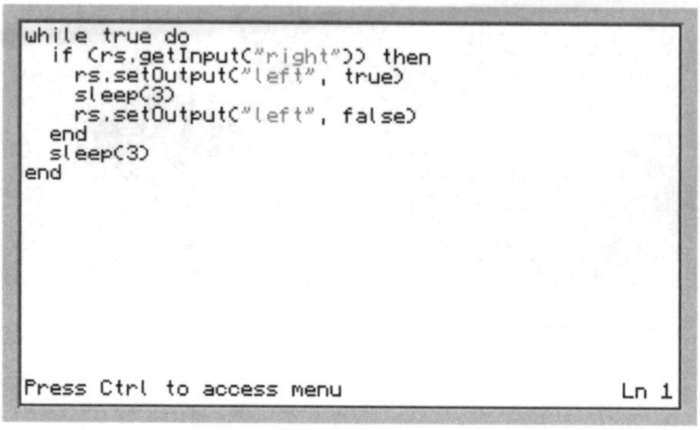

```
while true do
  if (rs.getInput("right")) then
    rs.setOutput("left", true)
    sleep(3)
    rs.setOutput("left", false)
  end
  sleep(3)
end




Press Ctrl to access menu                    Ln 1
```

Figure 8-7 *Editing the program*

Press ESCAPE to return to the game and flip the lever to turn the power on. You should see that the redstone lamp slowly blinks. When you have had enough and want to stop the program running, return to the computer and hold down CTRL and T for a second or so and the program will terminate, returning you to the terminal.

Example Code

All the Lua code used in this chapter is available from the book's website www.minecraftmastery.com and its github repository here: https://github .com/simonmonk/minecraftmastery/tree/master/computercraft_ examples

You can copy the files into a ComputerCraft computer's disk by finding the location of the drive.

A shortcoming of ComputerCraft's editor is that it does not allow you to cut and paste code from outside of Minecraft. For the small programs we have been writing up to now, this isn't usually a problem. However, as the programs get bigger, it would be nice to be able to get access to the program files using our PC or Mac's file system.

Fortunately for us, every time you save a program in ComputerCraft, the program file is also saved on your hard disk. However, it is a little tricky to track these down. The following instructions assume that you have installed ComputerCraft as part of Tekkit. The program files will be hidden away in your Technic Launcher folder.

From here, navigate to modpacks/tekkitmain/saves/. You will see a list of your Tekkit worlds here. Select the one you are playing in and then navigate to "computer." Here you will find folders numbered from 0 upward. There will be one for each of the computers in your world. Within that folder you will find all your Lua programs, which you can just open using whatever text editor you prefer.

You will also notice a "disk" folder in here, where you can also find files on your floppy disk.

Let's now examine the code behind this. First of all, we want the flashing to continue indefinitely, so we put all the code that we want to run into a **while** loop. All the lines between **while true do** and **end** on the last line will be run over and over again automatically.

The first thing that we encounter inside the **while** loop is an **if** statement. This is here because we only want to flash if the lever to the right of the computer is on. So, after the word **if**, we have a condition that must be true for us to carry out the statements between **then** and the first **end**. In this case, the condition is:

```
if (rs.getInput("right")) then
```

The condition, highlighted in bold in the preceding code, must be contained within parentheses. The command **rs.getInput** is a bit like the command **rs.setOutput**, except that rather than setting a redstone output (say to turn on a lamp), it checks the input, in this case the input on the right-hand side of the computer. Only if this is on are the statements between "then" and the first **end** run.

These statements set the output to the left to be true (as we did earlier from the command line). The command **sleep(3)** tells the computer to go to sleep for three seconds. We then have another **setOutput** command, but this time, it sets the left output to false, turning the lamp off again.

Outside of the **if** we have another **sleep** command, which will be run whether or not the input is turned on. This may seem a little odd, but one of the quirks of the computer is that when it is running a program, it goes to sleep for at least some of the time, to give the rest of Minecraft time to refresh and do other things before it wakes up again.

If you move this **sleep** command to be inside the **if**, then when the lever is not active, the program will whiz around and around without any chance for anything else to happen, and after a short while you will get the error message "Flash.lua: 2: Too long without yielding".

The sleep period that we used is "3", which means three seconds between turning the output on and off. If we want the lamp to blink faster, then we need to change this value in both places where we currently have **sleep(3)**. You do not have to restrict yourself to whole seconds, so try editing flash.lua again, changing both sleep periods to "0.5" (half a second). You should now find that the lamp blinks much more quickly.

A Light Chaser Example

You can use all sides except the front of the computer to control redstone output, so let's build an example that flashes four redstone lamps in sequence. Build the contraption shown in Figure 8-8.

The lamps to the left and right of the computer will be powered directly without any intervening redstone. The bottom of the computer is connected to the third lamp from the left using redstone wiring in a trench. The back of the computer is connected to the rightmost lamp behind the blocks.

Now switch to the computer's terminal by right-clicking on it and start the editor on a new file, which we will call "chaser.lua". Enter the text as

Figure 8-8 *A light chaser*

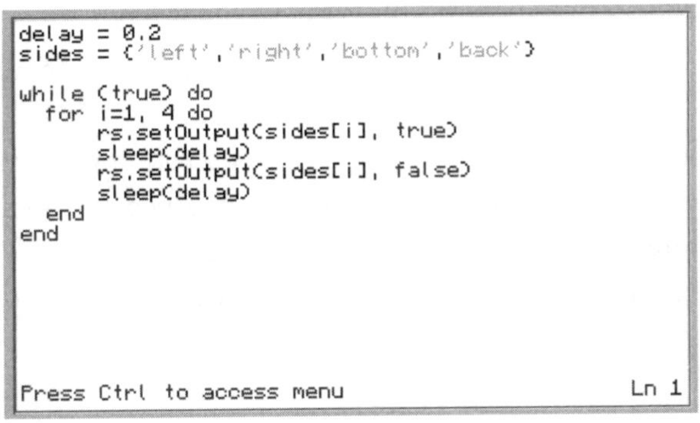

```
delay = 0.2
sides = {'left','right','bottom','back'}

while (true) do
   for i=1, 4 do
      rs.setOutput(sides[i], true)
      sleep(delay)
      rs.setOutput(sides[i], false)
      sleep(delay)
   end
end



Press Ctrl to access menu                    Ln 1
```

Figure 8-9 *Editing chaser.lua*

shown in Figure 8-9 and then save the file using CTRL then s, followed by CTRL and then E to exit the editor.

Run the program using the command:

```
>chaser.lua
```

This program is a little more complicated than the last one, but not much.

One idea that we do use in "chaser.lua" is that of variables. Variables are names that we give to a value, and that value can vary. So, for example, we have given the variable *delay* the value 0.2 in the line:

```
delay = 0.2
```

If you look through the text, you will see that later on, we use this name in the sleep commands:

```
sleep(delay)
```

This has the advantage that if we wanted to change the speed at which the lamps chase along the line, we only need to change the value of *delay* in one place. For example, to make the thing much slower, we could change the delay to two seconds by changing the first line to:

```
delay = 2
```

Variables can be used for collections of values as well as single values like *delay*. To simplify the process of looping around four of the sides in turn, we have used a list of the sides, and assigned them to the variable *sides*:

```
sides = {'left','right','bottom','back'}
```

The values in the list are text rather than numbers. Text like this is called a *string* in programming. In Lua, strings can be enclosed by single or double quotes.

This is another program that we want to keep running forever, and so all the lines after the variables are defined are contained inside a **while** loop. Inside this **while** loop is another type of loop called a **for** loop. The following line effectively counts from 1 to 4, setting the variable *i* to each of 1, 2, 3, 4 in turn:

```
for i=1, 4 do
```

If you remember, the command **rs.setOutput** expects two values to be given to it inside the "(" and ")" separated by a comma. The first value is the side to use (top, bottom, left, right, or back) and the second is "true" or "false" for on or off.

As we count through the four sides in turn, we want to use the side taken from the list of sides held in the variable *sides*. We can retrieve a particular item from this list by using square-bracket notation. So, **sides[1]** refers to the first element of sides ("left") and **sides[4]** to the last element "back". The command **sides[i]** will fetch the value from the list for the *i*-th side in the list. This will loop around from "left" to "back" as *i* changes from 1 to 4.

Computer Networks

ComputerCraft will actually allow you to create more than one computer in a world and what is more, to connect the computers together in a network. To accomplish this, you just need to add a couple of computers and attach a wireless modem to the sides of them. Yes, ComputerCraft even has wireless modems!

As an example, let's create an arrangement of two computers, each with a wireless modem on the side. One computer (the controlling computer) will have a lever on one side. The other will be placed on a small tower and have a redstone lamp on top of it. Build up the arrangement as shown in Figure 8-10.

To attach the modem to the side of the computers, you need to use "stealth" or the computer console will just open. So, to attach the wireless modem, have it in your hand and then hold SHIFT down while you right-click on the

Figure 8-10   *A wireless network in ComputerCraft*

computer. Install the program shown in Figure 8-11 on the controlling computer with the lever and that of Figure 8-12 on the computer with the redstone lamp on top.

Once both the programs are running, when the lever is switched, the lamp on the remote computer will light. You could of course do this with redstone wiring, but building it this way means that it can work without wires. You can also build some other identical receiving computers, running the same program and they will all respond to signals from the controlling computer, turning their lamps on and off at the same time.

The code for this is all pretty straightforward. Referring to Figure 8-11 for the sending code, you can see that the commands concerned with networking are all prefixed with "rednet". The first of these, on line 1, opens a connection to a wireless modem. The modem is attached to the left of the computer, hence "left" is passed in parentheses to the **open** command.

```
rednet.open("left")
while true do
  if (rs.getInput("right")) then
    rednet.broadcast("on")
  else
    rednet.broadcast("off")
  end
  sleep(0.1)
end




Press Ctrl to access menu                    Ln 1
```

Figure 8-11   *The program for the controlling computer*

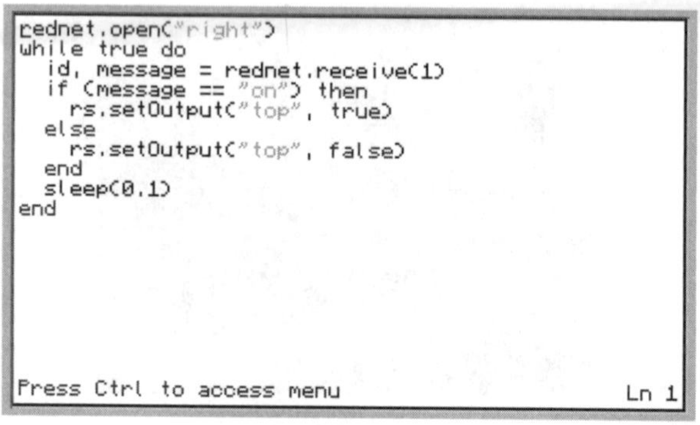

```
rednet.open("right")
while true do
  id, message = rednet.receive(1)
  if (message == "on") then
    rs.setOutput("top", true)
  else
    rs.setOutput("top", false)
  end
  sleep(0.1)
end




Press Ctrl to access menu                          Ln 1
```

Figure 8-12   *The program for the lamp computer*

Within the **while** loop, we check to see if the lever is on, using the following command:

```
if (rs.getInput("right")) then
```

If it is, then we broadcast the message "on" to any computers that might be listening on the network. Alternatively, if the lever is not in the "on" position, then we send the message "off". Finally, we sleep for 0.1 of a second to give the computer a chance to get on with other things.

The code at the receiving end is fairly similar (Figure 8-12). It also opens a connection to its modem and then jumps right into a **while** loop. Here it checks for any incoming messages using the following lines:

```
id, message = rednet,receive(1)
```

This will set two variables, *id* and *message*. The variable *id* is the identity of the sending computer. We do not use this in the program, but you still have to specify it. The second variable (*message*) will contain any message text that has been received through the wireless network. This is then compared to see if it is on, in which case it turns the top redstone output on, lighting the redstone lamp. Otherwise, it turns the output off.

Disk Drives

In the previous example, where we might be setting up a whole load of computers with the same program, it would save a lot of time if we could transport the program from one computer to another.

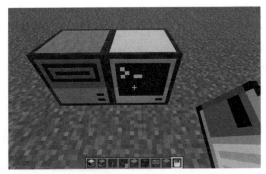

Figure 8-13   *A computer with disk drive*

You can do just this if you attach a disk drive to both computers. Figure 8-13 shows a computer with a disk drive attached to its left side.

In addition to a disk drive, you will also need one or more floppy disks to put in the drive.

Let's use a floppy disk to transfer a program file from one computer to another. First, insert a floppy into the drive, by right-clicking on the drive. When you do this, you will see something like Figure 8-14.

To insert the floppy disk, move it from your inventory into the "Disk Drive" square at the top. Press ESC to return to the game and then right-click on the computer, so that we can copy the program file onto the floppy, before ejecting it, taking it to a second computer with a disk drive, and copying the file back off the floppy. The copying takes place on the command line, as shown in Figure 8-15.

The first command (**dir**) will list all the files on the computer's file system. This shows two entries in green (**disk** and **rom**) followed by the file that we want to copy onto the floppy disk ("lamp.lua"). Linux and Mac users might like to know that you can use **ls** instead of **dir**.

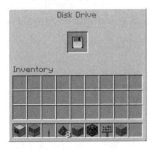

Figure 8-14   *Inserting a disk drive*

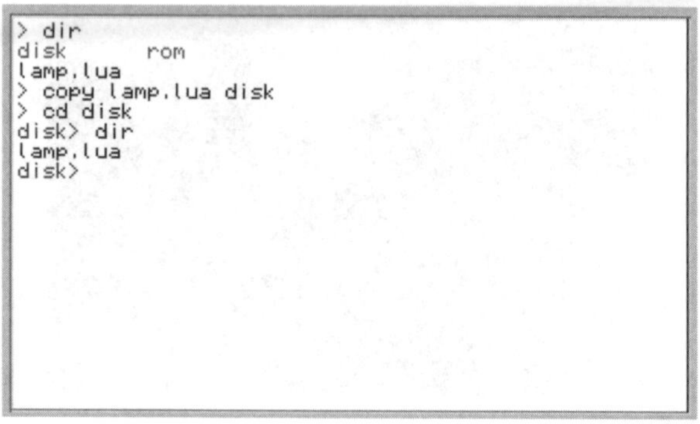

```
> dir
disk        rom
lamp.lua
> copy lamp.lua disk
> cd disk
disk> dir
lamp.lua
disk>
```

Figure 8-15   *Copying a file onto a floppy disk*

To copy the file, use the command:

```
> copy lamp.lua disk
```

In this case **disk** acts like a folder, so the **copy** command makes a copy of the file "lamp.lua" onto the floppy disk.

To check that the file is definitely on the floppy disk, we need to change directory to the disk drive using the command:

```
> cd disk
```

Note how the prompt has changed from ">" to "disk>" to remind us that we are looking at the contents of the floppy disk. Now when we enter the command **dir**, you can see that the disk does indeed contain the file that we want.

You can now eject the disk, by right-clicking the disk drive and dragging the floppy disk out of the top-central area (Figure 8-14) back into the inventory area. Move to a second computer with a disk drive attached and right-click on this second computer. The steps for copying the file off the floppy disk and onto the built-in hard drive of this computer are shown in Figure 8-16.

Change directory to the disk using the command:

```
> cd disk
```

You can now check that the file is there using the **dir** command. To copy the file from the floppy disk onto the built-in hard drive of the computer, use the command:

```
disk> copy lamp.lua /
```

```
> dir
disk    rom
> cd disk
disk> dir
lamp.lua
disk> copy lamp.lua /
disk> cd /
> dir
disk        rom
lamp.lua
>
```

Figure 8-16 *Copying a file from a floppy disk*

The "/" symbol refers to the root of the computer's file system. To confirm the copy, we can use **cd /** to change directory to the root and then **dir** to see the copied file in the file listing.

Having copied a file, you can if you wish destroy the disk drive. You can always attach a new one at a later date. It is probably a good idea to keep the floppy disk. You could use it to keep all your Lua programs on.

Monitors

Another useful feature of ComputerCraft is the ability to connect monitor screens to a computer and then display messages on them. Figure 8-17 shows a computer with a monitor attached.

As you can see, the monitor is of the widescreen variety. It is actually four blocks wide by two high. You can make the monitor as large as you like by simply adding more monitor blocks to it. However, it must be rectangular.

Figure 8-17 *Computer with monitor attached*

There is also a disk drive to the left of the computer, but this is not essential for the operation of the monitor; it is just there so that we can save the program to it if need be.

To display output on the monitor, you use the **monitor** command followed by the side of the computer that the monitor is adjacent to ("right" in this case) and then the name of the program that you want to run.

To illustrate the monitor in action, we can write a short program that just prints three lines of text (Figure 8-18). The program is called "message.lua".

The command **print** will display the string passed to it as a line of text on the monitor, automatically moving on to the next line after each **print** command. When you write to a monitor, the text stays on the screen even after the program has finished running, which is why there is no need to use a **while** loop in this example to keep the program running.

To run the program on the monitor, enter the following command in the computer console:

```
> monitor right message.lua
```

The preceding example writes all its output to the monitor specified when the program is run. It is also possible to attach multiple monitors to a computer and to send different output to each one. In the next section we will attach two monitors to the same computer.

You can find out more about monitors here: http://computercraft.info/wiki/Monitor.

Figure 8-18 *A test program for the monitor*

Figure 8-19 *A Minecraft clock*

Clock Example

This example (Figure 8-19) uses two monitors, one above the computer and one below it. Both are five blocks wide. The top monitor is three blocks high and the bottom monitor one block high. In this example both the computer and the monitors are of the Advanced variety. This allows them to display in different colors.

The code for this example is quite lengthy, so you may want to copy the file from the book's download page (www.minecraftmastery.com). The program is called "clock.lua".

```
bottomMon = peripheral.wrap("bottom")
bottomMon.setTextScale(2)
topMon = peripheral.wrap("top")
topMon.setTextScale(5)
topMon.setBackgroundColor(colors.white)
topMon.setTextColor(colors.green)

bottomMon.clear()
bottomMon.write("Approx Minecraft Time")

hour = 8 -- dawn
min = 0

while true do
  min = min + 1
  if min > 59 then
    min = 0
    hour = hour + 1
  end
```

```
if hour > 23 then
    hour = 0
end
topMon.clear()
topMon.setCursorPos(4, 2)
msg = string.format("%02d:%02d", hour, min)
topMon.write(msg)
sleep(1)
end
```

Because we are using two monitors, we cannot run the program using the usual **monitor** command. Instead, we run the program normally by just typing **clock.lua** at the computer's console. We connect to the monitors by using the **peripheral.wrap** command to assign each of the monitors (top and bottom) to a variable, like this:

```
bottomMon = peripheral.wrap("bottom")
```

Now, when we want to write something to the monitor, we can just use a command like:

```
bottomMon.write("Approx Minecraft Time")
```

The second line of the program sets the text scale for the bottom monitor to be 2. You can set this to anything from 1 to 5 to control the size of the text. The next few lines set up the top monitor in the same way; however, the background and text colors of the monitor are set to white and green respectively by the following commands:

```
topMon.setBackgroundColor(colors.white)
topMon.setTextColor(colors.green)
```

Two variables (*hour* and *min*) are used to hold the time of day. The hour is initialized to 8, which we will take to be Minecraft time 0 (that is dawn). We need a **while** loop, because we are going to continually update the display. So, each time, we add one to the *min* variable and then check that it is not greater than 59. If it is, it is set back to zero and one added to the hour.

The *hour* variable is checked in a similar way. When it exceeds 23, it is set back to zero. As you can see this is a 24-hour clock.

The next section of code clears the current contents of the display and then formats the hours and minutes into a text string before writing them to the display.

```
topMon.clear()
topMon.setCursorPos(4, 2)
msg = string.format("%02d:%02d", hour, min)
topMon.write(msg)
```

You can find out more about string formatting in Lua here: http://lua-users.org/wiki/StringLibraryTutorial.

Finally, there is a sleep of one second, as one second in real time corresponds to one minute of Minecraft time.

The clock will not keep very good time, as it takes no account of the time needed to refresh the display.

When working with ComputerCraft displays, you may find that sometimes the display does not display what you were expecting, or displays it in the wrong place. If this happens, then leaving the game and coming back in usually does the trick.

Turtles

ComputerCraft also includes turtles; these are essentially computer-controlled mobs. They are programmable just like a ComputerCraft computer. You can use turtles for digging, felling, mining, farming, and even melee. You can even attach a wireless modem to them to control them remotely.

Turtles have their own Application Programming Interface (API) that allows you to control what they do from the Lua program that they run. The commands in this API allow you to:

- Move the turtle using the commands **forward, back, up, down, turnLeft**, and **turnRight**.
- Dig using the commands **dig, digUp**, and **digDown**.
- Interact with the inventory using the commands **select, getItemCount**, and **getItemSpace**.
- Interact with blocks using commands including **place, detect, compare, drop**, and **suck**.

For a full list of these commands, see the API documentation at: http://computercraft.info/wiki/Turtle_(API).

Turtles require fuel to operate. That is, the turtle must have something that you would normally burn in a furnace, like coal, charcoal, or wood. You can change a configuration file to remove the need to refuel the turtle. The file is called "mod_CCTurtle.cfg" and you will find it in your technic directory technic/modpacks/tekkitmain/config/ (see "Installing the Example Worlds" at the start of this chapter to find the folder).

Open the file in an editor and search for the section that looks like this:

```
general {
    # Enable hardcore mode (turtles require fuel to
move)
    B:turtlesNeedFuel=true
}
```

Change the value "true" to "false".

Perhaps the simplest thing we can do with a turtle is have it do a little dance.

Since the turtle has an inventory to interact with, as well as being programmable, when you right-click on a turtle, you see the screen shown in Figure 8-20.

This particular turtle has 63 lumps of coal to power it. The main area of the window shows the program "dance.lua", which has just been saved to the turtle and can then be run using the command:

```
>dance.lua
```

Stand back a little before you run the program, as you may get in the turtle's way when it tries to move.

The first line of the program tells the turtle to refuel using slot 1 (the slot containing the coal) to a level of 10. If you omit this second parameter, then

Figure 8-20   *The turtle control screen*

the turtle will convert all the coal into fuel. But fixing it at 10 will have the advantage that the turtle will quickly run out of fuel.

Note that if you edit the "mod_CCTurtle.cfg" config file as described earlier, you can leave out this first line.

The turtle then goes into a loop, moving forward three blocks and then turning left. It will repeat this until it runs out of fuel or you catch up with the turtle, right-click it, and hold down CTRL-T. When the turtle runs out of fuel, it will still make the left turns, but it will not move forward. In other words it will just spin on the spot, without moving.

Let's now create a slightly more complex program that will lay redstone for us. To do this, we need to add some redstone next to the coal. Since the program will be quite similar to "dance.lua", copy the program using the following command:

```
> copy dance.lua redstone.lua
```

Figure 8-21 shows the program listing for this.

The second line selects item 2 (redstone) from the turtle's inventory. Inside the **while** loop, the turtle first moves forward and then turns left twice to turn through 180 degrees. This is so that it places the redstone behind it, so that it is not in the way when it tries to move forward again. The command **turtle.place()** places the redstone and then the turtle turns right twice so that it is facing forward again.

You can see the effect of this redstone laying in Figure 8-22.

You can adapt this program for all sorts of building tasks.

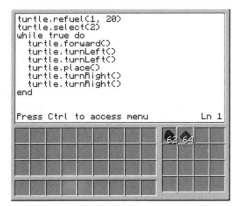

Figure 8-21 *A redstone-laying turtle*

Figure 8-22   *Laying redstone using a turtle*

Summary

In this chapter we have scratched the surface of the ComputerCraft mod. You can find out much more about this mod here: http://computercraft .info/wiki/Turtle.

In the next chapter we will look at how you can go about creating your own mods, with a little Java programming.

9

Modding with Forge

Minecraft is a very rich game, with many different types of block to play with. However, inevitably whatever is provided is still not enough for some people. This desire to do more with Minecraft has led to a whole community of "mods" like ComputerCraft and qCraft. While these are both examples of large-scale mods that involved a great deal of work to bring them to fruition, it is perfectly possible to make your own simple mods, in the form of different block types, or modifying existing blocks and items.

Indeed, if you take to Java, and have the time, you can create great mods for the Minecraft community to share.

Overview of the Modding Process

Minecraft is written in the Java programming language. So to write your own mods, you will need to learn some Java. You will need to install Java and also an IDE (Integrated Development Environment) in which to edit and debug the Java code that you write. The free and popular Eclipse IDE is the most frequently used IDE for modding, and so we will use this here.

Forge is the most popular framework for creating your own mods. It is a collection of Java code that simplifies the process of writing a mod. Mod writing is an unofficial process. That is, it is not officially sanctioned or supported by Mojang, the creators of Minecraft; however, they do not seem to mind, as ultimately you will still need to buy Minecraft to make use of the mods.

You can write mods on a Linux or Mac computer, but realistically, most people will be using some version of Windows. In fact, you should note that

139

it is not currently possible to write mods for platforms such as game consoles or the Raspberry Pi or mobile versions of Minecraft. This chapter describes the process of mod writing using Windows.

This chapter assumes that you are writing mods for version 1.7.2 of Minecraft. This version introduced some significant changes over earlier versions that will hopefully remain the same at least for a while, before these instructions become out of date. So, if you are working from a different version of Minecraft, then please refer back to the main Forge page (http://www.minecraftforge.net/) for the latest news.

Preparing Your Computer

Writing mods requires some fairly heavyweight Java programming tools, so you will need a decent computer to do this, or you will find yourself waiting quite a long time. Java works best with a lot of memory. As a guide, if Minecraft runs nice and quickly on your machine, then you should be okay writing mods.

Install Minecraft

It is unlikely that you have gotten this far into the book without installing Minecraft. However, you should keep a separate copy of Minecraft just for modding. So download Minecraft again from Minecraft.net and run it once and log in to create the necessary directories. There is no need to create a world; just get as far as the launcher.

Installing Java

While Linux and Mac come with Java preinstalled, Windows does not. So, your first step should be to install Java. This can in itself be confusing as there are many different versions and flavors of Java. You need to install a recent version of the JDK (Java Development Kit). Note that if you are distributing mods, then people using a later version of Java than the one you built the mod class with will get errors. We have used the process described in this chapter to build mods using versions 6 and 7 of Java. At the time of writing Java 8 has just become available. This may be fine, but we will assume the use of Java 7 in this chapter. If you are a Windows XP user, then Java 7 is the last version of Java available to you (Java 8 will not install).

Note that your PC may well have a "runtime" version of Java installed. This is not sufficient; you actually need the JDK.

You will find the downloads for the JDK on Oracle's website http://www .oracle.com/technetwork/java/javase/downloads/. Figure 9-1 shows the Java download page with Java SE (Standard Edition) highlighted.

After you click the Download button, you will be given the option of selecting the download for your operating system (Figure 9-2).

If you have a fairly new Windows computer, it will be 64-bit. So, select the file for "Windows-x64.exe". If you know that you have a 32-bit computer, then select "Windows x86". You will need to accept the license agreement before you can download the file.

The Java download is an executable installer, which then appears to install a second installer! Run it and the installation process will begin. You can accept all the defaults during the installation process and just watch the layers of installation window to make sure that you are clicking Next when you need to.

Adding Java to Your Path

Windows uses a system variable called PATH to search for programs to run. To make Windows aware of the version of Java that we have just installed, you need to add its location to the PATH variable. You can find full instructions for this here: http://www.java.com/en/download/help/path.xml.

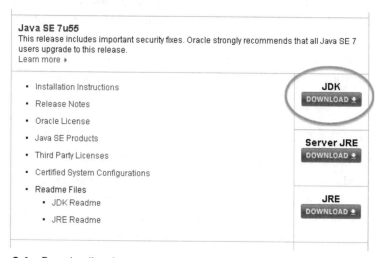

Figure 9-1 *Downloading Java*

Java SE Development Kit 7u55		
You must accept the Oracle Binary Code License Agreement for Java SE to download this software.		
Thank you for accepting the Oracle Binary Code License Agreement for Java SE; you may now download this software.		
Product / File Description	File Size	Download
Linux x86	115.67 MB	⬇ jdk-7u55-linux-i586.rpm
Linux x86	133 MB	⬇ jdk-7u55-linux-i586.tar.gz
Linux x64	116.97 MB	⬇ jdk-7u55-linux-x64.rpm
Linux x64	131.82 MB	⬇ jdk-7u55-linux-x64.tar.gz
Mac OS X x64	179.56 MB	⬇ jdk-7u55-macosx-x64.dmg
Solaris x86 (SVR4 package)	138.86 MB	⬇ jdk-7u55-solaris-i586.tar.Z
Solaris x86	95.14 MB	⬇ jdk-7u55-solaris-i586.tar.gz
Solaris SPARC	98.18 MB	⬇ jdk-7u55-solaris-sparc.tar.gz
Solaris SPARC 64-bit (SVR4 package)	24 MB	⬇ jdk-7u55-solaris-sparcv9.tar.Z
Solaris SPARC 64-bit	18.34 MB	⬇ jdk-7u55-solaris-sparcv9.tar.gz
Solaris x64 (SVR4 package)	24.55 MB	⬇ jdk-7u55-solaris-x64.tar.Z
Solaris x64	16.25 MB	⬇ jdk-7u55-solaris-x64.tar.gz
Windows x86	123.67 MB	⬇ jdk-7u55-windows-i586.exe
Windows x64	125.49 MB	⬇ jdk-7u55-windows-x64.exe

Figure 9-2   *Java SE download options*

In summary, to do this, open the Windows Control Panel and then select System and then Advanced. Click the button at the bottom that says Environment Variables, then select Path and click Edit (Figure 9-3). If there is no environment variable called PATH, you will have to create one. Assuming there is already a PATH variable, scroll back to the start of the "Variable value:" field and type **C:\Program Files\Java\jdk1.7.0_5\bin;** at the beginning of the field. Note that if you are using a later version of Java, you will need to change this path to agree with the location where you installed Java. Rather confusingly, Java 7 is also known as Java 1.7.

At this point, it is probably a good idea to reboot Windows so that the new environment variable takes effect.

Installing Eclipse

Go to http://www.eclipse.org/downloads/ to download Eclipse. Eclipse can be used as an IDE for many different programming languages, and the developers have helpfully provided configurations that are ready-made for use with various languages. Select the download to suit your architecture from the option Eclipse for Java Developers (Figure 9-4).

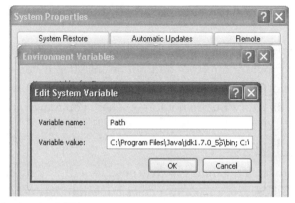

Figure 9-3 *Setting the PATH environment variable*

Figure 9-4 *Selecting the right version of Eclipse*

You will be asked for an optional donation and then the download will begin.

Once the download is complete, unzip the file. Note that this unzips the actual Eclipse program directory and not an installer. If you unzipped this into your directory, then you might want to move it somewhere else like your Program Files directory. But do not forget where you move it to.

Installing Forge

We now need to fetch the things we will need to start writing our own mods in Java using Eclipse. So, go to http://files.minecraftforge.net/, find the row for "1.7.2 latest", and click the link for "src".

Unzip the file to some convenient location such as the Desktop and rename the rather long directory name to just "forge".

Next you need to run the command prompt (also known as DOS prompt) from that directory. You can do that by clicking the Start menu and then Run, and then enter **cmd** in the field and click OK.

You now need to change directory to the location of the "src" download that you just unzipped by entering the following command at the command prompt:

```
>cd Desktop\forge
```

Note that the ">" indicates the command prompt; you do not need to type this. Now that you are in the forge source folder, you need to run a utility script that will set up an Eclipse workspace for you. Type the following commands to do this:

```
> gradlew setupDevWorkspace
> gradlew eclipse
```

The script will start producing a whole load of trace while it does its work. Eventually it will return to the command prompt, and this may take several minutes.

Setting Up Eclipse

Now it's time to run Eclipse. So go to the Eclipse directory that you downloaded and unzipped back in the section "Installing Eclipse." Double-click on the file eclipse.exe (a purple ball). You might also want to make a desktop shortcut for future use.

As Eclipse starts up, it will prompt you for the location of a "Workspace." Navigate to the eclipse directory within your forge directory.

Eclipse will allow us to make changes to our mods and then immediately run Minecraft from within the Eclipse environment to see how they work and debug them. To run Minecraft from Eclipse, click the Run button on the toolbar. If all is well, this will run Minecraft (Figure 9-5).

If this does not work, then your best option for fixing it is to search for the text of any error message using Google and look for other people who have had the same problems.

Figure 9-5 *Running Minecraft from Eclipse*

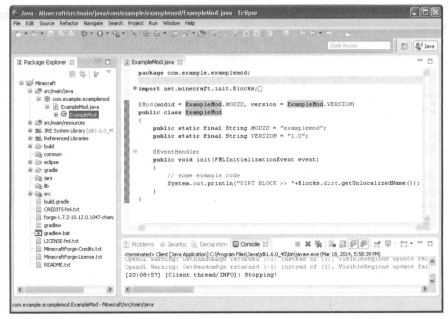

Figure 9-6   *The Eclipse IDE*

A QuickTour of Eclipse

Figure 9-6 shows the Eclipse IDE window. On the left-hand side there is a tree-type view of all the files involved in the project. If you click on one of the files and it is editable, then it will open up in the central editor area. This is where you will write the Java needed for your mods.

At the bottom of the window is the Console area where messages (especially error messages) will appear.

Java Basics

Before we start coding, let's look at some of the basics of Java programming. This book is nowhere near big enough to teach you Java programming; for that, there are many other very good books available. However, you can learn enough to get by when writing fairly simple mods without having to learn everything about Java.

Packages

When you look at the project organizer on the left of Eclipse, you can see a foldable tree structure of files. Java files are organized into "packages." These packages are really just folders inside folders. They follow a different naming convention when you use them in Java. They use a dot to separate the parts of the directory structure.

The package structure itself follows a convention, sometimes called the "reverse domain name." So, for example, the code for this book will be placed into a package called "com.minecraftmastery.thorium". In the next chapter, we will develop a mod called "thorium".

If we were to write a second mod, we might put it in a package called "com.minecraftmastery.cryptonite".

If you look at the start of any Java file, you will find a line something like:

```
package com.minecraftmastery.thorium;
```

This specifies the package that this Java file belongs to, and the package path must match the directory structure.

Underneath this line, you will find a load of lines starting with **import**. These specify other Java code (maybe in other packages) that this code needs in order to work. For example:

```
import net.minecraft.block.Block;
```

The first part of the import path is the package (**net.minecraft.block**), and the part after the final dot is what needs importing from that package. If you put a "*" after the package name, then everything from that package will be imported. If you are more specific, as is the case here, then you may just include one Java class (see the next section). Importing single classes is the safest option as it prevents problems if more than one package has a class of the same name.

Classes

Java code is organized into what are called "classes," and generally you will find that each Java file (of source code) will contain just one class. The class will have a name beginning with an uppercase letter and the file should have the same name, but with the extension of ".java" on the end of it. If we were to make a new type of block, then we would keep all the code relating to that block in a class. We might call that class "ThoriumBlock".

After the **package** and **import** statements, the following example shows how a class is defined.

```
public class ThoriumBlock {
}
```

Note that the name of the class must exactly match the name of the file but without the ".java" extension. The curly braces "{" and "}" are important as they delimit what is inside the class. In fact, Java uses curly braces in many different situations in order to group code together into "blocks" of code.

In this case the class is empty. We will need to add things between the curly braces later.

Member Variables

A class is a way of grouping together program code and data that relate to each other. A class will have one or both of those things. Data is kept in something called "member variables." An example of this would be:

```
public class ThoriumBlock {
  public Block thoriumOreBlock;
}
```

The Thorium block now has a member variable associated with it called *thoriumOreBlock* of type Block.

Methods

Program code (instructions to the computer on how to do things) is kept in "methods." These are also kept inside a class definition and belong to the class, like member variables. For example:

```
public void load(FMLInitializationEvent event)
{
    thoriumOreBlock = new ThoriumOreBlock();
    GameRegistry.registerBlock(thoriumOreBlock,
      ThoriumOreBlock.BLOCK_NAME);
}
```

Classes also often have a special type of method that has the same name as the class. This is the only time a method name should begin with an uppercase letter. These methods are called *constructors* and have the purpose of initializing a class after an instance of the class has been created.

Classes and Instances

A class is a one-off. You can think of it as a definition. An "instance," on the other hand, is an object of that type. So there will only ever be one class (of a particular name), but you might create one or more instances of that class using the **new** command.

If this all sounds a bit confusing, don't worry—you will see examples of this in the next chapter that should make things clearer.

Inheritance

When you create a class, you can optionally specify that it will have a parent class. So for example, if you create a class called "ThoriumOreBlock", it might need to have a parent class of "Block", imported from the package "org.minecraft.block.Block".

```
class ThoriumOreBlock extends Block
```

By using the **extends** keyword to specify a parent class like this, the new class "MyBlock" will "inherit" everything that "Block" has. That is, some of the methods and variables in "Block" will be available to use in "ThoriumOreBlock" without having to write a single line of code.

A common way of writing programs in Java is to define a general-purpose class like "Block" in the knowledge that someone is going to inherit from it. Because you can think of the "subclass," as it is called, as being a specialized version of its parent class, the writer of the general-purpose parent class might expect certain methods in the subclass to replace or override their more general counterparts in the parent class.

This is why you will often see a line like this preceding a method definition that replaces the method of the same name in the parent class:

```
@override
```

This tells the compiler, which turns your source code into something that will run, that the method definition must have the same method with the same parameters defined in the parent class; otherwise, it will be flagged as an error.

Public and Private

When you define a method (or member variable), you specify its accessibility; that is, whether code in other classes is allowed to use it. If a method falls into this category, it is a method that the programmer thinks might be useful to others, and it will be labeled as "public":

```
public void load(FMLInitializationEvent event)
```

If it is a method or variable that is just about the internal workings of that class, then it will be "private."

Naming Conventions

Some parts of the Java language are like laws—you have to stick to the syntax or things just won't compile. Other parts of Java are conventions that people follow simply to be good citizens. This makes it easier for other programmers to work out what the code is supposed to do.

Package naming in a reverse domain name format is one such convention. You do not have to stick to it, but there are good reasons why you should, especially if you wish to be considered a good programmer (as opposed to a "script jockey") by other programmers.

One convention is that class names should start with an uppercase letter and be bumpy case. That is, you start each new word with an uppercase letter. Here are some example class names:

```
Block
ThoriumOreBlock
Thorium
```

Similarly, variable names should start with a lowercase letter and also start each subsequent new word with an uppercase letter.

```
block
thoriumOreBlock
thorium
```

You will sometimes see the underscore (_) character used in variable names. This is a habit common in Python programmers, where it is the convention. It is not used much in Java, but like all rules, sometimes it is worth breaking and using an "_" to emphasize the importance of a variable.

Summary

In the next chapter, we will start to write out the first of two mods, starting with a block-based mod to get us started.

10

Example Mod: Thorium

This chapter is the first of two example chapters for modding. This first mod is based on the concept of a new type of block called Thorium. You will learn how to create an ore for this block that will be distributed about a world, ready to be mined. We will then develop the concept, so that when the ore is mined, it creates a Thorium block (Figure 10-1).

You will also learn how to use the new block in crafting recipes.

Setting Up the Project

If you have not already done so, read through Chapter 9 and set up the various tools that you are going to need, including Java, Forge, and Eclipse. Once it is all installed and you have confirmed that you can run Minecraft, including the Forge example mod, directly from Eclipse, then you are ready.

All the source code for this mod is available at the book's website and github repository (accessible from www.minecraftmastery.com). So, you can either use the repository files as a reference and open them in github, pasting them into newly created files in your Eclipse project, or you can just download the entire project. The instructions for installing the whole project are as follows:

1. Using your browser, navigate to https://github.com/simonmonk/ minecraftmastery.

2. Click on java and then thorium.zip.

3. To actually download thorium.zip, you will need to click the Raw button to its right.

Figure 10-1   *Thorium*

4. Once you've downloaded thorium.zip, switch over to Eclipse and then select Import from the File menu and then Existing Projects.

5. Click Select Archive File, navigate to where you saved thorium.zip, and click Open.

6. Finally, click Finish in the Import wizard.

You should now find that the Thorium project has been added to the project explorer.

Creating a Project

To keep our mod code separate from the rest of the Minecraft code, we are going to put all of our code in a separate project within Eclipse. So, if you are following this process from scratch, rather than importing the finished code, from the File menu of Eclipse select the option New Java Project and enter a project name of **Thorium**.

You can then accept the rest of the defaults by clicking Finish. Now you should have a second folder-type icon in the Package Explorer section of Eclipse labeled "Thorium".

By default Eclipse will put all the source code files and packages that contain them into a folder called "src". We need to change this so that the Java files go into a folder called "src/main/java". So, right-click on src in the Package Explorer and delete it.

Then right-click on the Thorium project in the Package Explorer and select the option New and then Source Folder. Enter **src/main/java** as the source folder.

Creating a Package

As we discussed in Chapter 9, Java expects you to organize your code into folder-like structures called packages. So let's start by creating a new package for our new Thorium mod.

Right-click on the "src/main/java" icon in the Package Explorer and select the option New Package. Leave the "source folder" field as "Thorium/src/main/java" but set the "Name:" field to "com.minecraftmastery .thorium" and click Finish (Figure 10-2).

Linking the Projects

At the moment, our new project is not linked in any way to the original "Minecraft" project that Forge created for us, so we need to link our new project to the "Minecraft" project so that Minecraft can use it.

To do this, right-click on the "Thorium" project and select "properties" from the end of the list. Next, click Java Build Path on the left and then select the "Projects" path. Now click Add, and check Minecraft, and then click OK (Figure 10-3).

Click OK again to close the properties window.

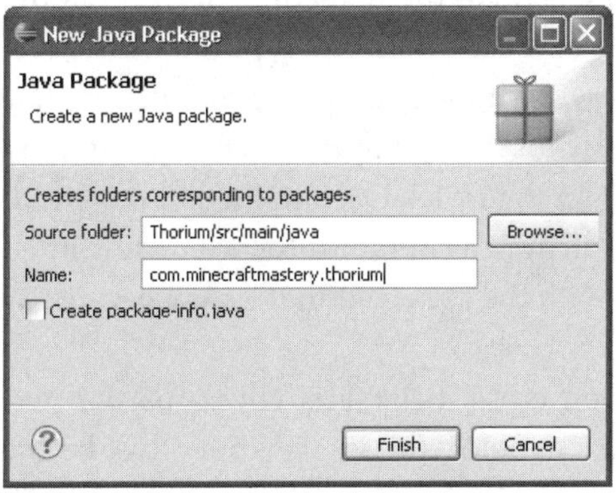

Figure 10-2 *Creating a new package*

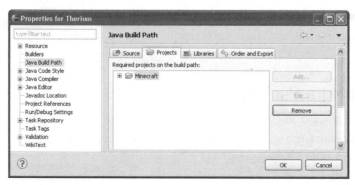

Figure 10-3   *Linking the projects*

Basic Block Ore

All the code that we will write is going to be placed into this package. We are going to write two classes. The first is a class to represent the Thorium ore and the second is to represent the mod as a whole (the "mod class"). When we add more blocks later on, each will have its own class, but also contribute some code to the shared "mod" class.

Creating the ThoriumOreBlock Class

Let's start by creating the class for our Thorium ore. Right-click on the package "com.minecraftmastery.thorium" and select the option New Class. In the dialog that appears, change the name to **ThoriumOreBlock** and then click Browse next to the Superclass field and type the word **Block** into the "Choose a Type" field. This will find all the classes that Eclipse knows about that are called "Block". We want the one in the package "net.minecraft.block" so select that (Figure 10-4), then click OK and then Finish to create the class.

The class code will appear in the editor with a few little red crosses. These indicate problems with the code, but that's okay because we haven't finished yet.

By showing a red cross, Eclipse is telling us that we need to define a constructor for our new block class. We know this, because if we hover the mouse over the red underlining on "ThoriumOreBlock," we will see the message shown in Figure 10-5.

Figure 10-4   *Creating the ThoriumOreBlock class*

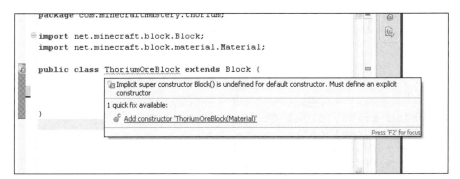

Figure 10-5   *Eclipse being helpful*

Packages and Eclipse

In the listings in this chapter, I have not displayed all the imports. In fact, Eclipse folds up the imports out of the way, just showing the first import in the editor. You can reveal any other imports by clicking on the little pointer next to the first **import** statement. Eclipse will also help you manage your imports. As you enter the text for this example into the Java file, you will sometimes be entering the name of a class that has not been imported yet. When this happens, the reference to the class name will have an error symbol next to it. If you click on this error, Eclipse will open a popup list from which you can select the class to import. This will automatically add it to the list of imports at the top of the file.

Pressing CTRL-SHIFT and o at the same time will automatically add imports to the file for everything you are using.

Let's do what Eclipse says and create a constructor. Type the extra text highlighted in bold in the following listing so that the whole class now looks like this:

```
package com.minecraftmastery.thorium;

import net.minecraft.block.Block;

import net.minecraft.block.material.Material;

import net.minecraft.creativetab.CreativeTabs;


public class ThoriumOreBlock extends Block {
  public static final String BLOCK_NAME =
  "thoriumoreblock";
  public ThoriumOreBlock() {
    super(Material.rock);

    setCreativeTab(CreativeTabs.tabBlock);
    setBlockName(BLOCK_NAME);
  }

}
```

We have added a string constant called **BLOCK_NAME** that we can use both inside the class, when calling the **setBlockName** method, but also from outside the class, as we shall see later.

You may notice that when you type the dot after **Material** or **CreativeTabs,** you will be presented with a list of possible completions. If you watch what Eclipse suggests to you, you can save yourself a whole load of typing.

Don't forget to save the file using CTRL-S or the Save option on the File menu.

Creating the Thorium Mod Class

Having created a class for the block itself, we can now create a class to represent the mod as a whole. To do this, right-click on the "com.minecraftmastery .thorium" package in the Package Explorer and select New Class. Enter a name of "**Thorium**." This time we do not need to specify a superclass, so just click Done when you have entered the name.

This will create an empty class that looks like this:

```
package com.minecraftmastery.thorium;
  public class Thorium {


}
```

We need to add some code here. Modify the class so that it appears as in the following example.

```
package com.minecraftmastery.thorium;

import com.example.examplemod.ExampleMod;

@Mod(modid = Thorium.MODID, version = Thorium.VERSION)
public class Thorium {
    public static final String MODID = "thorium";
    public static final String VERSION = "1.0";

    @EventHandler
    public void load(FMLInitializationEvent event)
    {
    }
}
```

We need some way of telling Forge that the class that we have created is for a mod, and is not just any old class. This is accomplished using the "@" annotation. You can think of this as adding labels to the class. In this case, it is identifying the class as being a mod and specifying a "modid" and "version" for the mod. The values for these labels are defined inside the class.

For example, the "public static final String" is called MODID. The "public" part means that it is visible to other classes. The word "static" means that it applies to the class as a whole rather than instances of the class. So, we can always get hold of this string using:

```
Thorium.MODID
```

You can see that we also have a method called "load" that is annotated as an @"EventHandler". This method will get run when Forge tries to load the mod. For now we will leave this method empty, but later on we will need to add some code into it.

Creating the mcmod.info File

Forge uses resource files so that when you run Minecraft, you can find out more information about the mods included. We need to create such a file for our mod, and it needs to be in a source folder separate from the Java source.

The first step is to right-click on the "Thorium" project in the Package Explorer and select the option New, then Source Folder. Enter a "Folder Name" of **src/main/resources**.

Next, right-click on the source folder you have just created and select the option New and then File. Specify a name for the file of "mcmod.info". Edit the file so that it appears as follows:

```
[
{
  "modid": "thorium",
  "name": "Thorium Mod",
  "description": "An example from the book Minecraft
Mastery.",
  "version": "1.0",
  "mcversion": "1.7",
  "url": "http://www.minecraftmastery.com",
  "updateUrl": "",
```

```
  "authors": ["Matt and Simon Monk"],
  "credits": "Wuppy - for his great videos that got us
started",
  "logoFile": "",
  "screenshots": [],
  "dependencies": []
}
]
```

You can change most of the fields to something shorter, but "modid" must match.

First Run

We have not done very much yet, but let's check our progress and make sure that we can run Minecraft with the mod, such as it is. To do this, we need to change the "run configuration" to be aware of our new mod, so click on the drop-down option next to the "Run triangle" and select the option Run Configurations. Here you need to click on Classpath and then User Entries and then Add Projects. Select Thorium and then click OK, then Apply, then Run. Having done this, next time we will be able to just click Run. This will start Minecraft running.

Click the Mods button and you should see the mod listed (Figure 10-6) along with its description.

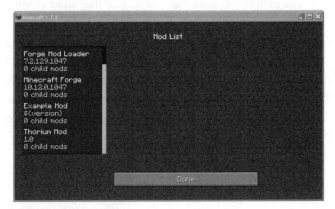

Figure 10-6 *The Thorium mod*

Making the Mod Usable

Although the Mod will show up in the list of mods, there are a few things we need to do before we can actually use it in a world. Modify the Thorium class so that it looks like this:

```
@Mod(modid = Thorium.MODID, version = Thorium.VERSION)
public class Thorium {
    public static final String MODID = "thorium";
    public static final String VERSION = "1.0";

    public static Block thoriumOreBlock;

    @EventHandler
    public void load(FMLInitializationEvent event)
    {
        thoriumOreBlock = new ThoriumOreBlock();
        GameRegistry.registerBlock(thoriumOreBlock,
          ThoriumOreBlock.BLOCK_NAME);
    }
}
```

The first thing to notice is that we have added a variable to the class:

```
public static Block thoriumOreBlock;
```

This variable will be used to store an instance of our ThoriumOreBlock class that will be created during the process of loading the mod. In fact, the code to create this instance is now the first line of the "load" method.

The second line of the method registers the block with the game, so that we will be able to find it in the inventory.

Now click the Run button again, but this time create a new world (select Creative mode). If you now go to your inventory and search for "Thorium"; you will find it there as a sort of purple checkerboard patterned block (Figure 10-7A). Add it to your inventory and you will find you can place the block in your world (Figure 10-7B).

A   B

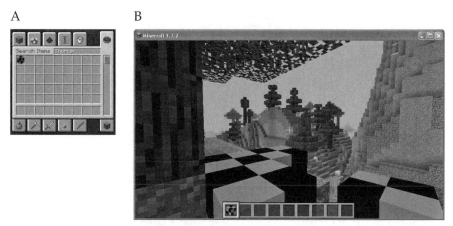

Figure 10-7   *The Thorium block in Minecraft*

Adding a Custom Texture

The appearance of the block in Figure 10-7 is the default appearance of any mod block. We now need to change the appearance of the block, to make it easily identifiable as Thorium.

Creating a Texture Image File

To give our block a custom appearance, we need to create an image file to use as the texture. The file for a texture needs to be 16×16 pixels and in PNG format. You will need an image editor such as GIMP (www.gimp.org). GIMP is available for pretty much any platform. Figure 10-8 shows the image I have drawn in GIMP to use as the texture for the Thorium ore block.

Because the image is so tiny, you will need to do a lot of zooming in to make it a reasonable size. To draw the image, you will also need to either find a very small brush or use the pencil tool. For interesting effects, you can also try scaling down a larger image to just 16×16.

Forge is very particular about where you put this image file for it to be available to Minecraft. We need to add a directory structure inside the "src/ main/resources" folder in our project. Right-click on "src/main/resources" in the Package Explorer and then select the option New and then Package. Enter a package name of **assets.thorium.textures.blocks**.

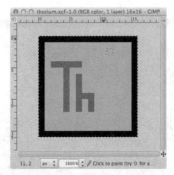

Figure 10-8 *Using GIMP to create a texture file*

Now take the PNG file you created, name it **thoriumoreblock.png**, and drag it from wherever it was created on your file system onto the package that you just created. You will be prompted as to whether you want to link to it or copy it into the project. Select the option to copy it. If you unfold the package in the Package Explorer, you will see the image there. You will also be able to click on the image to view it, but not unfortunately to edit it.

Finally, you need to register the icon with Minecraft, so add the following method to the class ThoriumOreBlock. This will override the version of the method in "Block".

```
@SideOnly(Side.CLIENT)

public void registerBlockIcons(IIconRegister reg) {
      blockIcon = reg.registerIcon(Thorium.MODID + ":" +
                 BLOCK_NAME);

}
```

Notice how we have joined together the string constants for both the MODID and BLOCK_NAME to create a string that will be "thorium:thoriumoreblock" to act as a unique identifier for the image.

Run Minecraft again from Eclipse and create a new world in Creative mode and then play it. Find Thorium in your inventory and you should find that it has the new appearance (Figure 10-9).

Localized Names

You may have noticed that if you hovered over the Thorium ore in your inventory, it will have named it something like "tile.thoriumoreblock.name"

Figure 10-9 *Thorium with a custom texture*

rather than simply "Thorium ore" as you would expect. This is because there is one step that we have not yet taken for our mod, and that is to define a language file. There may be different versions of this file containing translations of the block name for different languages. The only language translation that you really must provide is for U.S. English, and if no other translation files are provided, this will be used.

The language file must be created inside the assets package for the mod. So we will need to create a new package in "src/main/resources". Right-click on this folder in the package explorer, select the option New Package, and enter **assets.thorium.lang**.

Now right-click on the package and create a file called **en_US.lang**. This will open the empty file up in an editor. You now need to place the translation for our block name in there in the following format:

```
tile.thoriumoreblock.name=Thorium Ore
```

In the downloaded version, the files use item.thoriumblock.name; however, both seem to work.

Note that the first part of this is the message we got when hovering over the block in the inventory. Save the file and run Minecraft again. This time, when you hover over the block in your inventory, you will see the correct name.

If you create language files for other languages, these will automatically be switched in when Minecraft is run in another language. For example, you could create a French language file called "fr_FR.lang" containing this line:

```
tile.thoriumoreblock.name=Minerai Thorium
```

The first part of the file name "fr_FR.lang" is language and the part after the underscore is the country. The two are separated because languages often have different spellings and other variations between countries.

Spawning Ore Blocks

For the new block to be useful in Survival mode, we need it to be spawned at random locations when a new world is created, so that it can be mined.

When Minecraft creates a new world, Forge will ensure that all the mods will be given the opportunity to do any world creation stuff that they need to do. To do this, we will need a new "WorldGenerator" class in our mod. Let's call it **ThoriumWorldGenerator**.

Right-click on the "com.minecraftmastery.thorium" package and create a new class called "ThoriumWorldGenerator". Before you click Finish, click the Add button next to Interfaces and add **IWorldGenerator**. This will save you some typing. When the class code is generated, it should look like this:

```
package com.minecraftmastery.thorium;

import java.util.Random;

import net.minecraft.world.World;
import net.minecraft.world.chunk.IChunkProvider;
import cpw.mods.fml.common.IWorldGenerator;

public class ThoriumWorldGenerator implements
IWorldGenerator {

    @Override
    public void generate(Random random, int chunkX,
            int chunkZ, World world,
        IChunkProvider chunkGenerator,
          IChunkProvider chunkProvider) {
        // TODO Auto-generated method stub

    }

}
```

By specifying the Interface, Eclipse has automatically added in the "generate" method needed by Forge to add in our Thorium ore blocks. At least it has given us the skeleton of the method, and now we just need to fill in the details.

If we get world generation wrong, then we will crash Minecraft, so for the business of creating the actual blocks, I have taken a utility method described by "Wuppy" in his tutorials at http://www.wuppy29.com/minecraft/modding-tutorials/forge-modding/ and updated it for Minecraft 1.7.

This utility method should be added to the class ThoriumWorldGenerator. It is listed here with the comment block at its start removed for the sake of brevity. At this point, you will probably want to fetch the code for this from the book's website, rather than type it all in.

```
public void addOreSpawn(Block block, World world,
                        Random random, int blockXPos,
                        int blockZPos, int maxX,
                        int maxZ, int maxVeinSize,
                        int chancesToSpawn,
                        int minY, int maxY){
    int maxPossY = minY + (maxY - 1);
    assert maxY > minY: "The maximum Y must be greater
                        than the Minimum Y";
    assert maxX > 0 && maxX <= 16: "addOreSpawn: The
                        Maximum X must be greater than 0
                        and less than 16";
    assert minY > 0: "addOreSpawn: The Minimum Y must be
                        greater than 0";
    assert maxY < 256 && maxY > 0: "addOreSpawn: The
                        Maximum Y must be less than 256
                        but greater than 0";
    assert maxZ > 0 && maxZ <= 16: "addOreSpawn: The

                        Maximum Z must be greater than 0
                        and less than 16";

    int diffBtwnMinMaxY = maxY - minY;
    for(int x = 0; x < chancesToSpawn; x++)
    {
        int posX = blockXPos + random.nextInt(maxX);
        int posY = minY + random.nextInt(diffBtwnMinMaxY);
        int posZ = blockZPos + random.nextInt(maxZ);
        (new WorldGenMinable(block, maxVeinSize)).
                generate(world, random, posX, posY, posZ);
    }
}
```

What makes this method so useful is the **assert** statements at its start, which will check that those conditions are true, and if any are not, then it will produce an error message rather than continue in a way that would cause unpredictable results in Minecraft during world generation.

The actual core of the function that does the work starts after the **assert** statements and is quite concise. It repeats the lines inside the **for** loop the number of times set in the "chancesToSpawn" parameter, generating an ore block each time using random numbers constrained by the other parameters such as "maxY" that are passed to the method.

This method will not be called automatically. We need to add some code to "generate" to use it. So, modify the generate method so that it looks like this:

```
@Override
public void generate(Random random, int x, int z,
                     World world,
                     IChunkProvider chunkGenerator,
                     IChunkProvider chunkProvider) {

  if (world.provider.dimensionId == 0) {
    addOreSpawn(Thorium.thoriumOreBlock, world, random,
             x*16, z*16, 16, 16,
             4 + random.nextInt(3), 500, 50, 90);
  }
}
```

This will specify that when creating the surface world (dimensionId of 1), use the parameters supplied to generate blocks at random. The parameters are

- The block to spawn (ThoriumOreBlock)
- The world to spawn it in
- A reference to a random number generator to use
- The last block positions for X and Z
- The maximum X and Z coordinate values

- The maximum number of blocks in the vein

- The number of blocks to create per chunk

- The range of Y values (heights) at which the blocks can spawn

These parameters are set to create a lot of ore to make it easy to find during testing. You will want to reduce the Y range and the number of blocks per chunk down from 500 to perhaps 50.

The final step is to link our new world generator class to our mod class. So switch over to the class Thorium and add the following line to the class just after the line where you define a block:

```
public static ThoriumWorldGenerator thoriumWG;
```

Then add a line to "load" so that the load method is as shown in the following example.

```
@EventHandler
public void load(FMLInitializationEvent event)
{
    thoriumOreBlock = new ThoriumOreBlock();
    GameRegistry.registerBlock(thoriumOreBlock,
      ThoriumOreBlock.BLOCK_NAME);
    GameRegistry.registerWorldGenerator(thoriumWG, 1);
}
```

If you create a new world and go and visit it now, you should find it pretty easy to find some Thorium ore (Figure 10-10).

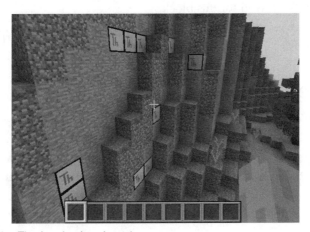

Figure 10-10   *Thorium in abundance!*

Mining Thorium

If you start a world in Creative mode with the code we have so far, you will find that when you try to mine the Thorium ore, it just breaks and there is nothing to pick up.

There are two methods that we must override in the ThoriumOreBlock class to make mining possible.

Add the following two methods to the class:

```
@Override
public int quantityDropped(Random random) {
    return 1;
}

@Override
public boolean canHarvestBlock(EntityPlayer
  player, int meta) {
    return true;
}
```

The quick way to add these methods is to right-click in the editor window, where you want the methods to appear, and then select the option Source, then select Override/Implement Methods and check the boxes for the two methods above. You will then need to edit the body of the methods so that they agree with the preceding code.

The first of these methods specifies that one block should be dropped during mining and the second simply specifies that "harvesting," or in this case mining, is allowed for this block.

Create a new world in Survival mode and you should now be able to mine Thorium ore.

Forging Thorium Ore into Thorium

We can take our Thorium example a stage further by defining a new type of block called just "Thorium" that is created by mining "Thorium ore" and then placing it on the crafting table to create Thorium. We will create a different icon for the Thorium and later on we can give it some other interesting properties.

Creating a New Block

Using the same method as you did for the Thorium ore, create a new class called **ThoriumBlock**. The code is listed here:

```
public class ThoriumBlock extends Block {

        public static final String BLOCK_NAME =
                "thoriumblock";

        public ThoriumBlock() {
                super(Material.rock);
                setCreativeTab(CreativeTabs.tabBlock);
                setBlockName(BLOCK_NAME);
        }

        @SideOnly(Side.CLIENT)
        public void registerBlockIcons(IIconRegister reg)
        {       blockIcon = reg.registerIcon(Thorium.MODID +
                        ":" + BLOCK_NAME);
        }

}
```

You will also need to create a texture file for the new block or use the one supplied in the downloaded code for the mod. This will need to be placed in the package "assets.thorium.textures.blocks" with the name "thoriumblock .png".

You will also need to modify the Thorium class, adding a new variable for the ThoriumBlock and registering it. The revised class is listed in the following example:

```
@Mod(modid = Thorium.MODID, version = Thorium.VERSION)
public class Thorium {
    public static final String MODID = "thorium";
    public static final String VERSION = "1.0";

    public static Block thoriumOreBlock;
    public static Block thoriumBlock;
    public static ThoriumWorldGenerator thoriumWG = new
                ThoriumWorldGenerator();

    @EventHandler
    public void load(FMLInitializationEvent event)
```

```
    {
          thoriumOreBlock = new ThoriumOreBlock();
          thoriumBlock = new ThoriumBlock();

          GameRegistry.registerBlock(thoriumOreBlock,
                    ThoriumOreBlock.BLOCK_NAME);
          GameRegistry.registerBlock(thoriumBlock,
                    ThoriumBlock.BLOCK_NAME);
          GameRegistry.registerWorldGenerator(thoriumWG, 0);
    }
}
```

Edit the language file en_US.lang, adding the following line to it:

```
tile.thoriumblock.name=Thorium
```

Once this is done, run Minecraft from Eclipse and check that both the Thorium and Thorium ore blocks can now be found in the inventory.

Adding a Crafting Recipe

Let's now define a recipe that will convert Thorium ore into Thorium blocks. We will just use a simple recipe where nine blocks of Thorium ore on the crafting table result in one block of Thorium (Figure 10-11).

To define the recipe, add the following method to the Thorium class:

```
private void addRecipes() {
      ItemStack thoriumStack = new
        ItemStack(thoriumBlock, 1);
      ItemStack thoriumOreStack = new
        ItemStack(thoriumOreBlock, 1);
```

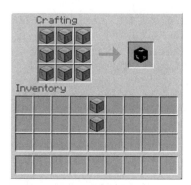

Figure 10-11   *Crafting Thorium*

```
Object[] params = new Object[] {
            "ooo",
            "ooo",
            "ooo",
            'o', thoriumOreStack};
GameRegistry.addShapedRecipe(thoriumStack,
                            params);
}
```

The first thing to note is that this method is declared as "private" because it is not intended to be used outside of this class. If you had other recipes to add, you would also put them in here.

To call the addRecipes method when the mod is loaded, you must add the following line to the end of the "load" method in the Thorium class:

```
addRecipes();
```

The recipe uses a pattern to indicate where in the crafting grid different types of block go. These are represented by characters, where a space means nothing required in that grid cell. In this case, I used the letter "o" to indicate the location of Thorium ore blocks, but any letter can be used and more than one letter used if different substances are required in the recipe. After the grid contents, the "params" object array has pairs of entries, each consisting of the letter used in the crafting grid (in this case, "o"), followed by an ItemStack containing the type of block (in this case, Thorium ore). A second ItemStack contains the results of the recipe cooking, in this case, a Thorium block.

Making Thorium Glow

There are a number of properties that can be changed for a block type when it is created. For example, you could make Thorium emit light by changing the constructor for ThoriumBlock to include a line to set the brightness. This is highlighted in the following code:

```
public ThoriumBlock() {
        super(Material.rock);
        setCreativeTab(CreativeTabs.tabBlock);
        setLightLevel(1.0f);
        setBlockName(BLOCK_NAME);
}
```

The light level is specified between 0 and 1.0 where 0 is no light and 1.0 is maximum. The letter "f" after the number indicates that the number is of type "float". There are other properties that you can set by starting a new line in the method, typing **set** and then pressing CTRL-SPACE. Other methods that you might want to use to set properties of the block are "setHardness", "setLightOpacity", and "setStepSound".

Summary

In this rather long chapter, we have explored a fairly simple example in considerable depth. You should now have a good understanding of what it takes to make a simple block-based mod. In the next chapter we will look at creating mods that use items but also how to share your mod creations rather than simply running them from Eclipse.

11

More Modding Examples

In this chapter, we will look at other aspects of modding that were not covered in Chapter 10. But first we will discover how to use the Thorium mod that we created so that we can run it without having to use Eclipse.

Releasing a Mod

So far, we have only been using our Thorium mod when running Minecraft from Eclipse rather than our normal Minecraft installation. This is the best way to do this while you are developing your mod, but when it comes time to share the mod with other people, this approach does not work.

Forge mods are distributed as a single Java Archive (JAR) file. This is a type of zip file, and if you have the JAR file for a mod, you can install it in a Minecraft with Forge simply by putting the JAR file in the mods folder of Forge.

This section will show you how to first create a JAR file, and then install the Forge Modloader on your normal Minecraft Launcher, and finally make use of your new mod.

Building a JAR File

Building the JAR file relies on the same build program (Gradle) that we used back in Chapter 9 to set up our Eclipse modding environment. However, this time, rather than use it to make an Eclipse workspace, we will use it to build the JAR file for a mod.

This build process is going to take place within the Thorium directory, so this time, we will not be able to run Gradle from the directory where it was installed. For this reason, we will actually install Gradle properly rather than use the bundled version that comes with Forge. This is pretty straightforward:

1. Navigate your browser to http://www.gradle.org/downloads and download the latest version of Gradle. I used version 1.11.

2. Save the folder somewhere convenient. Gradle runs from a folder; it does not have an installer.

3. Add the bin directory of the gradle folder you just downloaded to your Windows PATH variable, as you did when installing Java in Chapter 9. See Figure 9-4. If you installed Gradle on the desktop, then you might be adding something like this to the PATH variable: **C:\Documents and Settings\Simon\Desktop\gradle-1.11\bin**. Remember the semicolon to separate it from the other parts of the path.

Gradle doesn't do anything without a build file to tell it what to do. This is included in the Thorium project download, but let's assume that you are creating it from scratch, so in Eclipse, right-click on the Thorium project, select the option to create a new file, and name it **build.gradle**. This is at the top level under Thorium, not in any of the source packages.

The "build.gradle" file looks like this:

```
buildscript {
    repositories
    {
        mavenCentral()
        maven {
            name = "forge"
            url = "http://files.minecraftforge.net/maven"
        }
        maven {
            name = "sonatype"
            url = "https://oss.sonatype.org/content/repositories/snapshots/"
        }
    }
    dependencies
    {
        classpath "net.minecraftforge
                .gradle:ForgeGradle:1.1-SNAPSHOT"
    }
}
```

```
apply plugin: "forge"

sourceSets
{
    main
    {
        java { srcDirs = ["$projectDir/src/main/java"] }
        resources { srcDirs = ["$projectDir/src/main/resources"] }
    }
}

archivesBaseName = "thorium"
version = "1.0"
minecraft.version = "1.7.2-10.12.0.1047"
```

The most important bits are at the end of this file. First, the "java" and "resources" sections must match the locations of your java and resource files, which, relative to the Thorium folder where "build.gradle" is located, are in "src/main/java" and "src/main/resources."

The "archiveBaseName" should just be "thorium" and the value of "minecraft.version" must match the version you have been developing the mod with. You can find this by going to the Eclipse Package Explorer and opening the "Minecraft" project, in which you will also find a file called "build.gradle." Scroll down to the bottom of this file and copy and paste the "minecraft.version" from that file to be sure.

It's now time to run Gradle, so open a command prompt. Open the Windows Start menu, then click Run, then type **cmd** (on Windows XP). In the command prompt, change to the directory where you just created "build.gradle" and enter the following command:

```
>gradle build
```

If all is well, this will start the build process and after a while you should see the message "BUILD SUCCESSFUL" (see Figure 11-1).

You will find the JAR file itself inside the build/libs folder. This file contains everything needed to put your mod into a Minecraft installation.

Installing a Mod

To install the mod in Minecraft, we need to have a version of Minecraft that is:

- Compatible with the mod (in this case version 1.7.2)
- Has the Forge Modloader installed on it

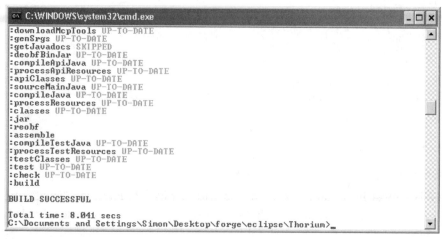

Figure 11-1 *Building the JAR file*

Fortunately for us, version 7.2 and up of Minecraft has a handy feature called "profiles" that allows you to easily run different versions of Minecraft from the same launcher. The installer for the Forge Modloader will also use this feature to create a "forge" profile. But first, we need to have run Minecraft once at the correct Minecraft version (1.7.2) to create the directory in which Forge can do its magic.

So, if you have not already done so, fetch a version of Minecraft, 1.7.2 or later, and run the launcher. Click Edit Profile and change the Minecraft version to "release 1.7.2" as shown in Figure 11-2.

The eagle-eyed among you may have noticed that the screen capture of Figure 11-2 is taken from a Mac. To demonstrate the portability of our mod, we will take the mod, developed on a Windows computer, and install it on Minecraft running on a Mac. The process of installing the mod is much the same whatever the platform.

We now need to add the Forge Modloader to Minecraft. To do this, visit http://files.minecraftforge.net and click the Installer link on the row for 1.7.2-Latest in the "Promotions" section of the download.

This will download a JAR file that is an installer for the Forge Modloader. When it has downloaded, double-click it to run it. It should pick up your Minecraft folder automatically. If it hasn't, check that you have run Minecraft 1.7.2 once. Figures 11-3A and 11-3B show the installer on Windows and Mac respectively.

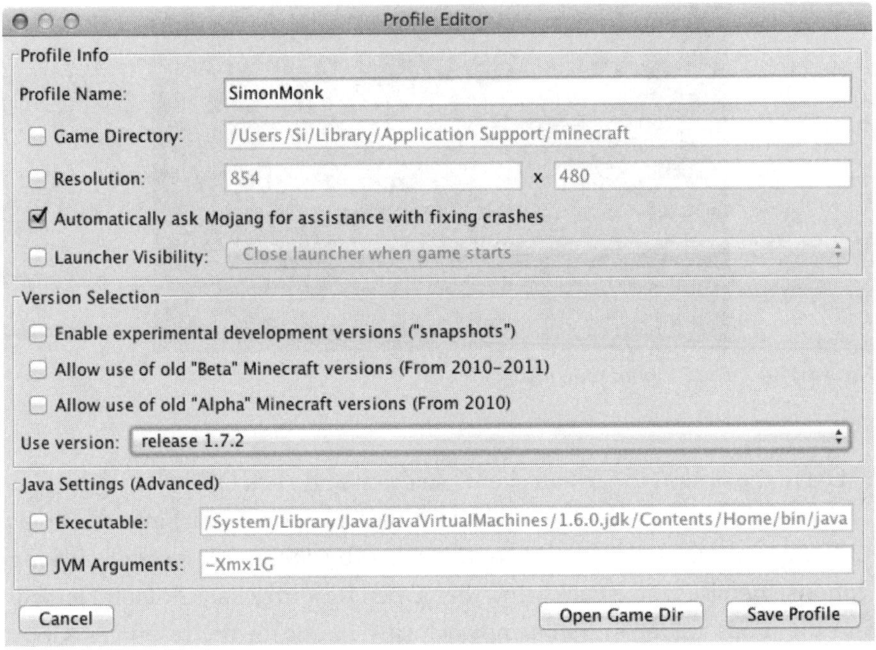

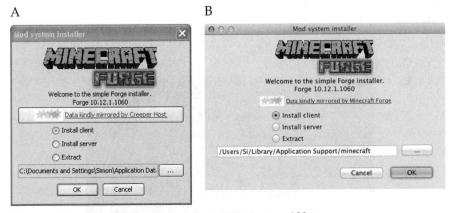

Figure 11-2 *Changing to version 1.7.2*

A   B

Figure 11-3 *Forge Modloader installer on Windows and Mac*

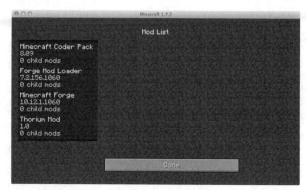

Figure 11-4 *The Thorium mod installed*

Leave the option to "Install client" selected and click OK.

You now need to copy the mod's JAR file (thorium-1.0.jar) into the mods folder. You can find the location for this folder by running Minecraft, clicking Options, then Resource Packs, and then Open Resource Pack Folder. You will find the mods folder in that same enclosing folder for the resources. Copy thorium-1.0.jar into this mods folder and you are good to go.

Run Minecraft again and this time you should have a new profile called "forge." Select this profile and click Play. Click the Mods button and you should see Thorium there (Figure 11-4).

Start a new world and mine yourself some Thorium ore (Figure 11-5).

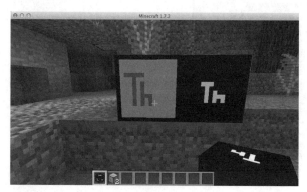

Figure 11-5 *Thorium*

More on Modding

The Thorium mod illustrates the process of creating different types of block in a mod, but does nothing in the way of items. In this section we will develop a second example mod that explores some of the other things you can do with a mod. The remainder of this chapter is all concerned with the "SmallThings" example mod.

In the following sections, we will be explaining how this example mod works rather than leading you through creating it as we did in Chapter 10, so you should download the SmallThings mod, and install it into Eclipse to follow the description.

All the source code for this mod is available at the book's website and github repository (accessible from www.minecraftmastery.com). The instructions for installing the whole "SmallThings" project are as follows:

1. Using your browser, navigate to https://github.com/simonmonk/ minecraftmastery.

2. Click on "java" and then "smallthings.zip."

3. To actually download "smallthings.zip," you will need to click the Raw button to its right.

4. Once you have downloaded the file, switch over to Eclipse, and then select Import from the File menu and click Existing Projects.

5. Click Select Archive File, navigate to where you saved "smallthings .zip" and click Open.

6. Finally, click Finish in the Import wizard.

You should now find that the "SmallThings" project has been added to the Project Explorer.

The SmallThings mod is a mixture of different item types, including armor designed to illustrate various points about mod making.

The mod follows the same pattern as the Thorium mod. These are some of the key files and folders in the project:

- **SmallThingsMod** The main mod file that coordinates all the other Items that make up the mod

- **src/main/java** A source folder containing all the Java code

- **src/main/resources**   A source folder containing all the icon images and language file

- **mcmod.info**   Information about the mod for display in the Minecraft Launcher

Take a few minutes to familiarize yourself with the layout of the project.

Item Example: Cooked Bone

Let's start by creating a variation of bone called "cooked bone." The idea is that by cooking regular bone (Figure 11-6), you will be able to make the new Cooked Bone material that has a damage boost effect, making the player who eats it stronger and more able to defeat mobs and other players in combat.

This new item requires a class of its own, and this is called "ItemCookedBone." It extends the class "ItemFood" as it is something that can be eaten. Its constructor is listed in the following example:

```
public ItemCookedBone(int p1, float p2, boolean p3) {
          super(p1, p2, p3);
          setPotionEffect(Potion.damageBoost.id, 5, 0, 1.0F);
          setCreativeTab(CreativeTabs.tabFood);
          setUnlocalizedName(ITEM_NAME);
}
```

After calling its parent class constructor with the same parameters as it received, it then sets a potion boost effect for the item. The remaining three parameters (5, 0, and 1.0) are the duration that the potion lasts, the effect level, and the probability of it having an effect. The constructor also sets the name and the inventory tab for the item.

If you go and look in the "SmallThingsMod" class, you will find that there is a member variable for this item:

```
public static Item cookedBone;
```

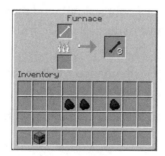

Figure 11-6   *Cooking bone to make cooked bone*

The following lines in the **load** method initialize the variable and add it to the game registry.

```
cookedBone = new ItemCookedBone(8, 0.8F, true);
GameRegistry.registerItem(cookedBone, ItemCookedBone.ITEM_NAME);
```

We do, of course, need a 16×16 PNG image for the item's icon, which you will find in "src/main/resources/" under "assets.smallthings.textures .items" and a language entry in the file "en_US".

As yet, there is nothing that tells Minecraft how to make the cooked bone. In fact, in this mod, the way we handle recipes in general is slightly different than in Chapter 10. In this mod, we use a separate class called "SmallThingsModRecipes" to contain all the recipes. This is not essential, but it helps to keep the **load** method in "SmallThingsMod" from becoming enormous.

The "SmallThingsModRecipes" class inherits from "SmallThingsMod," which means that it automatically has access to its member variables such as *cookedBone* that contain all the different types of item in this mod. All the work of defining the recipes is in the single method in "SmallThingsModRecipes" called **addRecipes**. The **addRecipes** method is called from the end of the **load** method in SmallThingsMod.

If you look at the top of this **addRecipes** method, you will see the following line:

```
GameRegistry.addSmelting(Items.bone, new ItemStack(cookedBone, 1), 0.2F);
```

This adds the smelting recipe to Minecraft. The first parameter is the item that is needed, in other words the "input." The second parameter is an ItemStack containing the result of the cooking and the quantity to be produced (the output). The final parameter (0.2) is the experience bonus to the player for doing this.

Item Example: Hardened Diamond

The second example of a new item in this mod is that of a Hardened Diamond. This item is only included to act as an ingredient for creating an obsidian axe in the next section.

As with the Cooked Bone example, the item is contained in its own class, in this case: ItemHardenedDiamond. Take a look at the code. You will find that it is almost identical to the Cooked Bone example item.

Item Example: Obsidian Axe

The Hardened Diamond item will be used in a crafting recipe for the obsidian axe that we will make next.

The class for this axe is called "ItemObsidianAxe." This item is pretty similar to the items we have met so far. One difference is that rather than having a parent class of "Item," this class uses the built-in "ItemAxe" class as its parent class.

```
public class ItemObsidianAxe extends ItemAxe{
```

This means that it will automatically inherit the abilities and properties of an axe. We just need to make sure that we supply it with the necessary image file in "assets.smallthings.textures.items" and text entry in the "en_US.lang" file.

If you look in the "SmallThingsMod" class, you will find this line that defines properties of a new tool material:

```
public static ToolMaterial toolObsidian =
    EnumHelper.addToolMaterial("OBSIDIAN", 3, 2561, 7.5F, 4.0F, 10);
```

This will be needed later when creating an ItemObsidianAxe to add to the game registry. The parameters are firstly the name of the material ("OBSIDIAN"), and then the harvest level (3), which relates to the number of blocks it can destroy. The next parameter (2561) is the durability of the tool itself. That is how soon it will wear out. The fourth parameter (7.5) is the speed at which it can destroy blocks; the fifth (4.0) is the damage the tool does during combat, measured in half hearts. The final parameter (10) is a bit of a mystery. It is known to be "enchantability," but it is not obvious how Minecraft uses this information, if it uses it at all.

This tool material is passed as a parameter to the constructor for ItemObsidianAxe when the member variable *obsidianAxe* is initialized in SmallThingsMod's **load** method.

```
obsidianAxe = new ItemObsidianAxe(toolObsidian);

GameRegistry.registerItem(obsidianAxe, ItemObsidianAxe.ITEM_NAME);
```

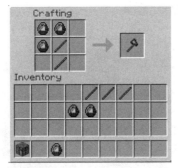

Figure 11-7 *Crafting an obsidian axe*

The constructor for ItemObsidianAxe simply passes this on to its parent class:

```
protected ItemObsidianAxe(ToolMaterial material) {
        super(material);
        setCreativeTab(CreativeTabs.tabTools);
        setUnlocalizedName(ITEM_NAME);
}
```

To craft one of three obsidian axes (Figure 11-7), we need to add a crafting recipe to SmallThingsModRecipes.

The code for this is similar to the crafting recipe for Thorium back in Chapter 10.

```
GameRegistry.addRecipe(new ItemStack(obsidianAxe),
      new Object[] {"HH ",
                    "HS ",
                    " S ",
                    'H', hardenedDiamond, 'S', Items.stick});
```

You can see how the pattern that should appear on the crafting table is set out in the code. Two token letters "H" and "S" are used as stand-ins for the item types Hardened Diamond and Stick.

You can access any standard items for use in recipes using the "Items" class. Try typing **Items.** in an editor window on Eclipse and you will see a long list of possible item types.

Armor Example

Creating armor is a bit of a special case when it comes to creating items, because, unlike the other items that we have looked at so far, you can wear armor (Figure 11-8).

Figure 11-8 *A player resplendent in obsidian armor*

Incidentally, if you want to be able to look at yourself in Minecraft, press the F5 key.

Armor Item Classes

To create a full set of armor, we need to create four new item classes: "ItemObsidianHelmet," "ItemObsidianChestplate," "ItemObsidianLeggings," and "ItemObsidianBoots." Each of these is on its own a fairly standard item, much like the ones that we have already looked at in this chapter.

If we look at one of these item classes, we can see things are a little different:

```
public class ItemObsidianHelmet extends ItemArmor{

    public static final String ITEM_NAME = "obsidianhelmet";

    public ItemObsidianHelmet(ArmorMaterial material, int p1, int p2) {
            super(material, p1, p2);
            setCreativeTab(CreativeTabs.tabCombat);
            setUnlocalizedName(ITEM_NAME);
    }
    public void registerIcons(IIconRegister reg){
        itemIcon = reg.registerIcon(SmallThingsMod.MODID + ":" + ITEM_NAME);
    }

    public String getArmorTexture(ItemStack stack, Entity entity,
            int slot, String type){
      return SmallThingsMod.MODID + ":textures/armor/obsidianArmor_1.png";
    }
}
```

In addition to the usual **registerIcons** method, which will link the item to its icon, we also have a second method called **getArmorTexture** that is responsible for returning a link to a separate image file that will be used to dress the player. There are two of these texture files defined (obsidianArmor_1 .png and obsidianArmor_2.png). The first of these is used when rendering any part of the armor except the leggings, which use the second of these files.

Armor Texture Files

These texture files are not the same as the usual 16×16 pixel files, but rather are a composite image containing various parts of the armor in different areas of the file. Two files are used, one for most of the armor parts and one for just the leggings.

Figures 11-9 and 11-10 show these two images.

One way to make images for new armor is to take these images from the project and edit them, retaining the same outlines for the different body parts. There will also normally be a fair amount of trial and error to get things appearing how you want them.

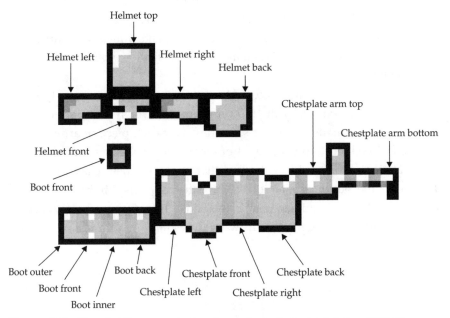

Figure 11-9 *The obsidian armor texture image for helmet, chestplate, and boots*

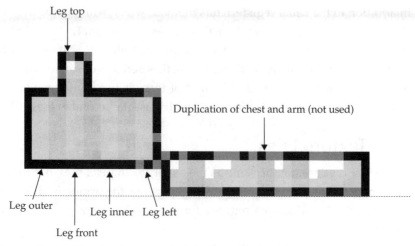

Figure 11-10   *The obsidian armor texture image for leggings*

These files are located in the src/main/resource source folder under "assets.smallthings.textures.armor".

Another way of making the armor textures is to use a website such as www.minecraft.novaskin.me. In addition to allowing you to make your own character skin, you can also use the website to make your own textures for anything in the game. It displays a model that will allow you to see how the armor will look as you edit it.

If you go to http://tinyurl.com/Armor-Editor-1, it allows you to draw over a diamond texture for the top three armor parts. If you right-click on each part individually, it will let you draw over each piece of armor. Once you've finished, go to the top-right of the screen, click Save and click Save again at the bottom right of the box that appears. Then, click Download Texture at the bottom center of the box that appears. Simply repeat the process with the leggings (using this link: http://tinyurl.com/Armor-Editor-2), and then you have your two textures for your armor.

Adding Armor to the Mod File

Turning our attention to the "SmallThingsMod" class, we need member variables for each of the Armor Item classes:

```
public static Item obsidianHelmet;
public static Item obsidianChestplate;
public static Item obsidianLeggings;
public static Item obsidianBoots;
```

We also need to add some lines to the **load** method to initialize them:

```
obsidianHelmet = new ItemObsidianHelmet(armorObsidian, 0, 0);
GameRegistry.registerItem(obsidianHelmet, ItemObsidianHelmet.ITEM_NAME);

obsidianChestplate = new ItemObsidianChestplate(armorObsidian, 0, 1);
GameRegistry.registerItem(obsidianChestplate, ItemObsidianChestplate.ITEM_NAME);

obsidianLeggings = new ItemObsidianLeggings(armorObsidian, 0, 2);
GameRegistry.registerItem(obsidianLeggings, ItemObsidianLeggings.ITEM_NAME);

obsidianBoots = new ItemObsidianBoots(armorObsidian, 0, 3);
GameRegistry.registerItem(obsidianBoots, ItemObsidianBoots.ITEM_NAME);
```

Notice that the last parameter to the constructor for each of the armor item classes changes. That is because this parameter determines the position of that armor element on the player's body. So, 0 is helmet, 1 is chestplate, 2 is leggings, and 3 is boots.

Armor Recipes

Finally, the crafting recipes for the armor items are all added to the class "SmallThingsModRecipes." These are all very similar to each other. The one for the helmet is shown in the following code:

```
GameRegistry.addRecipe(new ItemStack(obsidianHelmet),
    new Object[] {"HHH",
                  "H H",
                  'H', hardenedDiamond});
```

Graphical User Interfaces (GUIs)

When you right-click on a crafting table or a computer in ComputerCraft, a separate window opens up on top of Minecraft. You can do the same kind of thing in your own mods, where they are called GUIs. As an example of a GUI we will add a new block to the Tiny Things mod called "Switchable Light." When you right-click on one of these blocks, it will pause the game and open the GUI shown in Figure 11-11.

The idea is that you can position these Switchable Light blocks, for example, around your home, and then set them to either emit light or not emit light using the GUI.

Changes to the Mod Class

The first thing we need to add to the code is an "implements" clause to the mod class ("SmallThingsMod") so that it is able to handle requests to open a GUI. To do this, add "implements IGuiHandler" to the end of the class definition as shown here:

```
public class SmallThingsMod implements IGuiHandler{
```

When you do this, you will get errors telling you that to conform to that interface, the class has to implement two new methods, **getClientGuiElement** and **getServerGuiElement**. These two methods are shown following:

```
@Override
public Object getClientGuiElement(int ID, EntityPlayer player, World world,
            int x, int y, int z) {
    return new SmallThingsGUI();
}

@Override
public Object getServerGuiElement(int ID, EntityPlayer player, World world,
            int x, int y, int z) {
    return null;
}
```

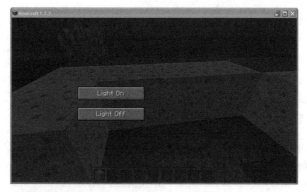

Figure 11-11 *A GUI with two buttons*

The first of these methods creates an instance of the GUI class that we will create in a moment. It then returns that newly created instance. The second method (**getServerGuiElement**) just returns null.

We also need some means of accessing the single instance of the Mod class ("SmallThingsMod") that will be created during the startup process of Minecraft and Forge. Normally, we do not need to access this instance, but in this case, we need to be able to bind the user interface back to this instance.

To do that, we define a new class variable on "SmallThingsMod" called INSTANCE:

```
public static SmallThingsMod INSTANCE;
```

Then in the **load** method of "SmallThingsMod" we assign "this" to INSTANCE:

```
INSTANCE = this;
```

To register this class as a GUI handler, we then need to follow this line in the **load** method with:

```
NetworkRegistry.INSTANCE.registerGuiHandler(this, this);
```

The "BlockSwitchableLight" Class

Most of the "BlockSwitchableLight" class is just the same as a regular block as described in Chapter 10. However, we do also override the method **onBlockActivated**.

```
@Override
public boolean onBlockActivated(World world, int x,
            int y, int z, EntityPlayer player,
            int i, float a, float b,
            float c) {
    player.openGui(SmallThingsMod.INSTANCE, 0, world, x, y, z);
    return true;
}
```

This method will be run every time a block of this type is right-clicked. It will be provided with a fairly comprehensive set of information about how this right-click happened, including the "world" object it occurred in, the player who did the clicking, and the coordinates. It does not, however, include the mod class responsible for the block. This is where the INSTANCE variable we added to SmallThingsMod comes in.

When the right-click occurs, we make the following call:

```
player.openGui(SmallThingsMod.INSTANCE, 0, world, x, y, z);
```

This informs the SmallThingsMod instance that we wish to open the GUI.

The "SmallThingsGUI" Class

In this example, the SmallThings mod only has one GUI, so this is not a bad name for the class. However, you can have more than one GUI in a mod, in which case SwitchableLightGUI would probably be a much better name for it.

The class inherits a load of useful things from GuiScreen, including knowing how to pause the game and display a greyed-out background while the GUI is displayed. It also inherits the behavior that when you press the ESCAPE key, the GUI is closed.

```
public class SmallThingsGUI extends GuiScreen {
```

The "SmallThingsGUI" class also has two variables to contain the two buttons of the interface:

```
GuiButton bOn = new GuiButton(1, 100, 100, 100, 20, "Light On");
GuiButton bOff = new GuiButton(2, 100, 130, 100, 20, "Light Off");
```

The parameters in the constructors for GuiButton are a button number (we don't actually use that), the x and y coordinates of the button, its width and height, all in pixels. The final parameter is the text to be displayed in the button.

We then need to override the **drawScreen** method.

```
@Override
public void drawScreen(int x, int y, float f) {
    drawDefaultBackground();
    super.drawScreen(x, y, f);
}
```

This calls the method on its parent "drawDefaultBackground" that does the graying out bit and then calls "drawScreen" on the parent class (that's what "super" means).

The buttons will not be automatically added to the GUI. To do that, we must implement the **initGui** method:

```
public void initGui()
{
        this.buttonList.add(bOn);
        this.buttonList.add(bOff);
}
```

This adds the buttons to a variable on the parent class ("GuiScreen") called *buttonList*.

The final piece of the jigsaw is to add some code to handle what happens when one of the buttons is clicked. This takes place in the **actionPerformed** method.

```
@Override
protected void actionPerformed(GuiButton button) {
        if (button == bOn){
                SmallThingsMod.switchableLight.setLightLevel(1.0F);
        } else if (button == bOff) {
                SmallThingsMod.switchableLight.setLightLevel(0.0F);
        }
}
```

This method will be called every time any of the buttons in the GUI are clicked. It supplied as its parameter the button that was clicked, so we can check to see which button was clicked and change the light level of the "switchableLight" to 1 or 0 appropriately.

There is a lot more to GUIs in Forge, but this should give you a start.

Summary

In this final chapter we have looked at a variety of code for modding items and also looked at how you can create a GUI that can be opened when you right-click on a block.

Appendix

Resources

Finding help on the technical aspects of Minecraft can be a frustrating business. In no small part, this lack of useful reliable information was a key reason for writing this book. The list in the following table is by no means complete, but will provide you with some useful tips.

URL	Description
minecraft.org	The official Minecraft site
minecraft.gamepedia.com	The Minecraft Wiki
qcraft.org	The official site for qCraft
computercraft.info	The official pages for the ComputerCraft mod
minecraftforge.net	The Forum for Forge. Tip: Search using [1.7.2] for information relating to modding with a particular Minecraft version.
minecraftforge.net/wiki/Wuppy's_tutorials	Wuppy's tutorial pages (written and video)
technicpack.net	The Technic Platform website, home of Tekkit. Loads of interesting mods available here.

When it comes to trying to follow the building of some complex contraption, video is a very good way of following the build. You will find many such videos on YouTube.

Index